Continental Scientific Drilling Program

U.S. Geodynamics Committee
Geophysics Research Board
Assembly of Mathematical and Physical Sciences
National Research Council

NATIONAL ACADEMY OF SCIENCES
Washington, D.C. 1979

The workshop on which this report is based was supported by the Department of Energy, the National Science Foundation, the Office of Naval Research, and the U.S. Geological Survey.

This work relates to Department of the Navy contract N00014-78-C-0501 issued by the Office of Naval Research. The United States government has a royalty-free license throughout the world in all copyrightable material contained herein.

International Standard Book Number 0-309-02872-8

Library of Congress Catalog Card Number 79-84872

Available from

Office of Publications
National Academy of Sciences
2101 Constitution Avenue, N.W.
Washington, D.C. 20418

Printed in the United States of America

Preface

The report *Continental Scientific Drilling Program* is the result of a Workshop on Continental Drilling for Scientific Purposes held in Los Alamos, New Mexico, July 17-21, 1978. It consists of a report of the U.S. Geodynamics Committee and six principal appendixes (A-F), which are the reports prepared by the panels of the workshop. Responsibility for the report itself rests with the U.S. Geodynamics Committee and the chairman of the workshop. The report is based largely on the findings and recommendations of the six panels.

A workshop on continental drilling held in 1974 led to the report *Continental Drilling* (1975). The U.S. Geodynamics Committee strongly supported the scientific recommendations in that report. The decision to hold the workshop in 1978 that led to the present report resulted in large part from the recognition that a substantial amount of drilling activity had developed in federal agencies. In April 1978 the U.S. Geodynamics Committee agreed to sponsor a Workshop on Continental Drilling for Scientific Purposes to be held in July 1978 and agreed that it should be so conducted that the report of the workshop would be completed by the end of the workshop.

Planning for the workshop was guided by a steering committee and six panel chairmen (Appendix G). To assist in achieving the goal of completing the workshop report by the end of the workshop, papers on pertinent topics were prepared and distributed to participants prior to the workshop (Appendix H).

Each panel report was prepared under the guidance of the panel chairman, who endeavored to take account of suggestions made during the meetings of the panels and to reflect a consensus of the discussions. There was considerable interaction among the panels during the course of the workshop, and some panel chairmen requested certain participants to serve as liaison members between the panels. Midway in the workshop, an afternoon was devoted to reviews of the conclusions in the draft report of each panel. In this way, all workshop participants had an opportunity to become acquainted with the directions being taken by all panels. The panel reports have subsequently been edited, but otherwise are reproduced essentially in the form in which they were submitted on the last day of the workshop.

Within the time constraints imposed on the workshop, the reports of the panels represent a broadly balanced view of the major scientific problems and concomitant needs for direct information obtainable from boreholes. However, because this constraint of time did not permit careful review of all panel reports by all participants, each individual participant may not necessarily agree with all statements in the panel reports.

The workshop was sponsored by the U.S. Geodynamics Committee with financial support of the Department of Energy, National Science Foundation, Office of Naval Research, and U.S. Geological Survey. By invitation of the Los Alamos Scientific Laboratory, the workshop was held in the National Security and Resources Study Center in Los Alamos, New Mexico, where it benefited from generous contributions of time by the staff of the Study Center and the staff of the Los Alamos Scientific Laboratory.

U.S. Geodynamics Committee*

JOHN C. MAXWELL, University of Texas (Austin), Chairman
CHARLES L. DRAKE, Dartmouth College
CLARENCE R. ALLEN, California Institute of Technology
DON L. ANDERSON, California Institute of Technology
ALBERT W. BALLY, Shell Oil Company, Houston
HUBERT L. BARNES, Pennsylvania State University
ARTHUR L. BOETTCHER, University of California
CARL KISSLINGER, University of Colorado
WILLIAM R. MUEHLBERGER, University of Texas
HARTMUT A. SPETZLER, University of Colorado
GEORGE A. THOMPSON, Stanford University
ROLAND E. VON HUENE, U.S. Geological Survey

ELBURT F. OSBORN, Chairman of the Workshop on Continental Drilling
for Scientific Purposes

AMPS-GRB Liaison Representatives

PRESTON CLOUD
ROBERT G. FLEAGLE
CARL H. SAVIT

*John C. Maxwell, Chairman from December 1978; Charles L. Drake, Chairman through November 1978; Clarence R. Allen, member through November 1978; Don L. Anderson, member from December 1978.

Reporters for the Special Topics* of the U.S. Geodynamics Committee

Topic		Reporter
1	Fine Structure of the Crust and Upper Mantle	JACK E. OLIVER
2	Evolution of Oceanic Lithosphere	JAMES R. HEIRTZLER
3a	Internal Processes and Properties	JOHN C. JAMIESON
3b	Crystal Growing	THOMAS M. USSELMAN
3c	Large Volume Experimentation	ROBERT E. RIECKER
4	Application of Isotope Geochemistry to Geodynamics	BRUCE R. DOE
5	Geodynamic Modeling	DONALD L. TURCOTTE
6	Drilling for Scientific Purposes	EUGENE M. SHOEMAKER
7	Magnetic Problems	CHARLES E. HELSLEY
8	Plate Boundaries	JOHN C. MAXWELL
9	Plate Interiors	LAURENCE L. SLOSS
10a	Geodynamic Syntheses	CREIGHTON A. BURK
10b	Data Centers and Repositories	ALAN H. SHAPLEY
11	Geodynamic Activities in the Caribbean Area	JOHN D. WEAVER
12	Lithospheric Properties	THOMAS H. JORDAN
13	Aeromagnetic Survey	WILLIAM J. HINZE

*Topics and reporters as of April 1978. The sequence of listing of these topics has no implication regarding priorities.

Contents

Background

Drilling for scientific purposes dates back many decades; the concept of a systematic drilling program began to evolve in the United States in the last twenty years. Reports of the Upper Mantle Committee in the 1960's, of the Geodynamics Committee in the 1970's, and others (see Bibliography) urged use of the drill to solve scientific problems where it is the appropriate tool and where the problems have been well defined. A number of countries have initiated drilling programs designed to respond to scientific needs.

In the United States a major scientific drilling effort began in 1968 with the Deep Sea Drilling Project (DSDP) and in 1975 with the International Phase of Ocean Drilling (IPOD). The deep sea was a logical place to begin, because targets of major scientific importance had been identified and further investigations using conventional techniques—without direct access to deeper subsurface levels that can be achieved only by drilling—were yielding only incremental increases in knowledge of the seafloor. The ocean science community strongly supported the project, had carefully considered problems for the drill to resolve, and was fully prepared to justify, plan, and manage such a program. The DSDP was initiated within the framework of existing technology but has made use of technological innovations. It was a program in which science led mission applications. The problems had been defined by exploration; drilling contributed to the solutions of these problems. Academic institutions provided much of the leadership in developing and conducting the DSDP. Industry and government agencies participated actively in the program and used the results almost as soon as they became available. The impact of the DSDP on ocean science, and on earth science as a whole, has been enormous.

At that time, the need for a coordinated drilling program on land was less evident. We now know that the continental crust evolved over a much longer period of time than that of the oceans, and it involved a greater number and diversity of processes. In comparison with marine geologists and geophysicists, the geoscience community concerned with the continents is larger, more diverse, and, historically, more focused on problems of local than regional or continental scale. Furthermore, major drilling efforts were being carried out by industry for resources and, to a lesser extent, by federal agencies in response to their missions. A few holes were drilled specifically for

1

scientific purposes by the academic community, and some efforts were made to use existing holes for downhole measurements.

In the 1960's a new concept—plate tectonics—was developed. According to this hypothesis, the outer shell of the earth is made up of a small number of large plates moving relative to each other and interacting at their boundaries. The DSDP made major contributions to verification of this model. The plate tectonics model was used with great success to explain the first-order features of the earth and the distribution of earthquakes and volcanic activity along plate margins. It has not had equal success in explaining the less frequent earthquakes and volcanoes within plates. It has been extraordinarily success-ful in explaining the nature and age of the ocean crust, less so the continental crust, particularly as age and depth increase. It has provided a powerful comprehension of the last 5 percent (ca. 200 m.y.) of earth history, but has been less effective in explaining or accommodating the 95 percent of earth history for which the repository is the continents.

Nevertheless, plate tectonics has provided two factors that strongly affect the rationale for a continental drilling program. First, it offered a framework to which geoscientists of all disciplines could relate their data. Thus it has been a highly effective unifying factor in the geoscience community and has led to productive multidisciplinary research. Second, it led to the realization that although the continental crust is complex, it is not impossibly so. There is order on a large scale that can be determined; this may in turn allow smaller-scale order to be recognized.

These two factors, plus the enhanced ability through application of new technology to explore the continental crust, provide a fresh rationale for a Continental Scientific Drilling Program. We can now select sites that will provide information about more than the immediate area of one borehole. We have the geological, geophysical, and geochemical tools to conduct the neces-sary exploration.

The continents do not stop at the shoreline, but extend to the continental slope or rim, under several thousand meters of ocean. Information on the water-covered portions of the continents is very important in resolving the evolution of the continents, and the drill can be expected to make significant contributions. The scientific opportunities in studying the modern conti-nental margins have not been emphasized in this report, as they have been covered in detail by IPOD, Joint Oceanographic Institutions for Deep Earth Sampling (JOIDES), Future Scientific Ocean Drilling (FUSOD), and the NAS-NRC study on continental margins. Ancient continental margins, subsequent-ly integrated geologically into the continental crust as we see it today, are considered in this report, and the results of drilling these ancient margins, combined with the results of offshore drilling of modern margins, should contribute greatly to our knowledge of their evolution through time.

The dynamics, structure, evolution, and genesis of the continents offer a major scientific challenge. The geoscience community responded to earlier challenges: the geographical challenge through exploration; the planetary challenge through spacecraft missions; and the challenge of the ocean depths by sampling from the ocean surface, by exploration with submersibles, and by drilling. Now the challenge is on land.

The Relation of Scientific Drilling to Proposed Geoscience Programs for the 1980's

The challenge of understanding continental evolution had led the International Union of Geodesy and Geophysics and the International Union of Geological Sciences to agree in principle that a well-designed international, interdisciplinary program in solid-earth science should be developed for implementation upon completion of the International Geodynamics Project in 1979. This decision was based on the remarkable scientific achievements under the present program and its predecessors and the clear perception of the importance of a continued international program involving the cooperative efforts of geologists, geophysicists, geochemists, and geodesists. The scientific program is currently under development, but it seems clear that the central focus will be the outer shell of the earth—the lithosphere—and that the dynamics of the crust will be a very important element of the program.

The U.S. Geodynamics Committee is in the process of developing a program of research for the 1980's. The development process involves close liaison and participation of the academic, industrial, and governmental geoscience communities. The committee is assisted by reports, current and pending, of many boards and committees of the National Research Council—boards and committees on seismology, geodesy, geochemistry, rock mechanics, ocean sciences, and mineral and energy resources—and by reports of special committees considering relevant scientific problems. It has cooperated with the Geophysics Study Committee of the Geophysics Research Board, whose studies on principles of mineral concentration (a joint study with the U.S. National Committee for Geochemistry), geophysical predictions, and continental tectonics are providing clarification of the promising directions in these areas.

The program of geodynamics in the 1980's under development by the U.S. Geodynamics Committee includes increased emphasis on continental crustal dynamics, especially as a framework for resource systems. It is based on the scientific knowledge gained during the International Geodynamics Project and its predecessors, the Upper Mantle Project (1962-1970) and the International Geophysical Year (1957-1959). The natural course of development of the geosciences has reached the point where they can significantly contribute to the understanding of the dynamics, structure, evolution, and genesis of the crust. At present, there is better understanding of the oceans than of the continents. Obtaining the desired knowledge of the continents will require application of all available tools and talents to the problem, and a finely tuned communication mechanism—one that will permit geoscientists of all disciplines to exchange their data and ideas and to incorporate information from areas outside their specialties to their own problems and results. Extensive geological and geophysical investigations have been made—and must continue to be made—on the surface of continents. The drill is the tool that enables direct measurement and sampling to be made at otherwise inaccessible depths below the earth's surface.

At present, mission-oriented drilling programs by governmental agencies are the primary means of probing the continental interior for scientific purposes. There are important possibilities for cooperation with industrial drilling programs, if funding can be arranged. As the scientific problems evolve and emerge, drilling efforts dedicated solely to seeking information on highest-priority scientific questions will also be appropriate. A long-term program must have flexibility to select among possible drilling arrangements.

Societal Problems, Agency Missions, and Drilling

Societal Problems—An Example

Society is faced with many problems whose solutions require the kind of information that can be obtained by a scientific drilling program—especially those associated with supplies of energy, water, and mineral resources, the mitigation of geological hazards and disposal of wastes, the effects of structures on the environment and of the environment on structures, and certain aspects of national defense.

The objectives of a continental drilling program for scientific purposes are described in the reports of four panels on basement structures, thermal regimes, mineral deposits, and earthquakes. The scientific results from such a program would be applicable to all of the problems mentioned above.

Nuclear waste isolation is an important example. No body of geological information that could be obtained from a scientific continental drilling program is more urgently required than criteria for site selection for isolation and management of radioactive wastes. The accumulated separate high-level wastes, the currently proposed spent-fuel storage, and the continuing accumulation of low-level wastes pose distinct waste-management problems. The alternatives of permanent versus retrievable storage or some optimum combination of these two requires the development of predictive capability for long-term site stability and assessments of the hazards of contamination of biological environments.

An analysis of the present state of geological information about various classes of sites must include, for any particular waste form, the following characteristics at depth in the earth's crust and associated factors:

- The chemistry and mineralogy of the selected medium
- The nature of crustal structure
- The existing state of stress
- The evidence for long-term site stability
- The physical response to stress perturbations (rock mechanics)
- The present local and regional hydrology and future hydrological changes
- The effects of water accessibility to the storage medium and to the waste form
- The kinetics of various thermal and chemical reactions to waste deployment
- The possibility of devising or selecting multiple, redundant forms of geological containment
- The nature and design of short- and long-term monitoring systems to measure the effectiveness of any operational site

The reports of the various panels of the Continental Drilling Workshop show that the primary scientific thrusts are directed toward developing measurement capabilities, a body of scientific data, and a conceptual understanding in all of the geological settings that have been proposed for waste isolation. Further, the breadth of the Continental Scientific Drilling Program offers potential for identifying both new classes of sites and specific sites.

The striking convergence of broad scientific interest with questions relevant to waste disposal is readily apparent. In connection with earthquake-hazard analysis, the Panel on Earthquakes focuses on learning how to measure the state of stress within the crust using drill holes. The Panel on Basement Structures discusses basement and deep-basin crustal structure, stability, and distribution of rock types. The Panel on Thermal Regimes is concerned with observation, measurement, and understanding of thermal perturbations in the crust and with hydrological permeability and interactions with hot rock. The Panel on Mineral Deposits is concerned with understanding the principles governing the migration and concentration of key elements in the crust, including many of the elements whose radioactive isotopes are formed as waste products. The Panel on Borehole Instrumentation identifies many problems that the scientific community currently faces in making accurate measurements of stress, permeability, rock chemistry, density, and other properties, especially at the high temperatures that might develop in a disposal site.

The panel discussions identify several areas in which there are critical gaps in our knowledge, techniques, and immediate capabilities for geological analysis of any proposed waste-isolation site. It is in the national interest to take immediate steps to improve the science base in areas where drilling is an essential technique.

Agency Missions and Drilling

Many federal agencies have missions that require knowledge of the dynamics, structure, evolution, and genesis of the continents, for the obvious reasons that we live and build our structures on them, extract our living and nonliving resources from them, and are subjected to the hazards of their instabilities. We are even able to alter the surface of the continent in geologically significant ways.

Agency missions are supported by general and specific appropriations, with planned schedules for implementation. The investments in specific drilling tasks may be many times larger than the operating budgets of the entire relevant scientific community for the same period. However, within a given agency's drilling programs, there are usually numerous opportunities for carrying out research of fundamental scientific importance. Therefore planning and coordination should be developed to take advantage of these opportunities and to minimize the incremental increases in costs and delays to the agency in completing its mission. To participate in cooperative efforts, the research community must be informed of possible opportunities, consulted for maximum yield to both the agency and itself, and funded to carry out its share of planning, preparation, and implementation of the desired research.

As agency missions require increasing national investments in drilling, more attention must be directed to overlaps, adequacy of design, and significance of results. The broad involvement of the scientific community in these drilling programs will help to assure that the national interest in an economical national drilling effort is served. Overlapping agency efforts that are unnecessarily duplicative can be avoided under the public-information mechanism that is proposed for the Continental Scientific Drilling Program. Similarly, the same mechanism may help identify significant gaps among various mission-directed drilling activities. Scientific and technical expertise available in one agency may be integrated into another agency's drilling effort. Interagency and intra-agency awareness of common interests and potential should be best served by such coordination.

Development of a Program

The U.S. Geodynamics Committee recognized the need for the kinds of data that could be obtained only by drilling. In its outline of the U.S. Program for the Geodynamics Project, it recommended that consideration be given to the contribution that could be made by continental drilling to resolving major problems of geodynamics. Subsequent activities led to a workshop supported by the Carnegie Institution of Washington held at Ghost Ranch, New Mexico, in 1974. The report from this workshop recommended a systematic program of continental drilling directed toward three problem areas: mechanisms of faulting and earthquakes, hydrothermal systems and active magma chambers, and the state and structure of the continental crust. The proposed program was designed to be responsive to the mission responsibilities of federal agencies, as well as to answer specific scientific questions. The report noted the relation of this effort to the increasing scarcity of mineral resources and suggested a framework for a national program of continental drilling.

The scientific aspects of the report were endorsed by the U.S. Geodynamics Committee, and the administrative aspects were carefully considered by the Federal Coordinating Council for Science, Engineering, and Technology (FCCSET) Committee on Solid Earth Sciences, which addressed the interests of many federal agencies in scientific drilling in connection with mission objectives. However, by the end of 1977 it became apparent that events had overtaken plans and that elements of a continental drilling program already existed in various federal agencies, particularly in the Department of Energy and the U.S. Geological Survey (of the Department of the

Interior). *The focus changed from one of initiating scientific drilling to one of maximizing the scientific value of current and planned efforts of federal agencies and industry and supplementing these efforts with holes drilled solely for scientific purposes.*

Federal agencies have evolved drilling activities calling for expenditures of more than $500 million per year; this might more accurately be termed the federal effort rather than a national drilling program. The drilling is in response to mission objectives; the scientific benefits are derivative, unlike the program of ocean drilling in which the science led the application. Given this new situation, it was considered timely to reexamine the conclusions of the earlier reports, to assess the scientific opportunities that the new situation presents, and to identify some of the scientific needs of the federal drilling effort. It was to accomplish this purpose that the Workshop on Continental Drilling for Scientific Purposes was convened.

A scientific drilling program should have three elements; the absence of any one of these seriously decreases its value. The first element comes before drilling; it is *problem definition*, which may involve investigations of many types, ranging from analysis of existing borehole data for geological investigations of geothermal sites to deep seismic reflection studies of scientifically interesting structures. The second element is the actual *drilling*, coring, and other sample collections together with the downhole measurements made during drilling operations. The third element is *postdrilling investigations*; they include studies of the cores, analysis, interpretation, and communication of the results and long-term monitoring in the holes. In general, actual drilling costs an average of between one third and two thirds of the cost of a total program of scientific drilling.

There may be great annual fluctuations in actual drilling costs, but the costs of the total scientific drilling program—advance exploration and problem definition, drilling and logging, and postdrilling investigations—should not fluctuate so widely. The three elements are inseparable and should be considered as a unit in budgeting for a scientific drilling program. Science is the program; the drill is the tool for that program.

Major Scientific Objectives

The design of a Continental Scientific Drilling Program must be predicated primarily on priorities developed from scientific considerations. The drilling program will be subject, of course, to legal, political, and economic constraints and will depend on opportunities to engage in cooperative drilling ventures, but the fundamental concern must relate to the scientific significance of the proposed effort.

The scientific objectives of the program are discussed under four headings: basement structures and deep continental basins, thermal regimes of the crust, mineral resources, and earthquakes and faulting. These headings are not all-inclusive, but they are responsive to the scientific directions for the 1980's identified by domestic and international scientific bodies, to agency missions and ongoing drilling programs, and to societal concerns.

In developing these programs the panels found much common ground and numerous opportunities for joint ventures to respond to several needs. *The degree to which cooperative ventures can be undertaken will depend strongly on a communication network designed to keep the entire geoscience community advised of future drilling plans.*

Basement Structures and Deep Continental Basins

The oceanic crust is created, ages, and returns to the earth's interior in geologically short periods of time. The continental crust appears to persist for a much longer period during which it is exposed to a variety of geological processes. The result is a complex body of rocks about whose state, composition, history, and distribution we know very little, except where they are exposed at the surface or where they have been explored for resources in the subsurface. Because we are highly dependent on the continental crust as a place to live and as the principal source of materials, it is not surprising that a better understanding of the continental crust—the composition, distribution, and temporal and structural relations of the rocks that constitute the crystalline basement of the continental United States, as well as the processes involved in their formation—is a prime objective of the U.S. earth scientists.

Adequate determination of the structure and evolution of the continental crust will be a formidable task. The drill is one of many tools available, but it is expensive; it must be used with particularly careful planning. The Panel on Basement Structures emphasized maximum use of holes of opportunity (holes drilled for specific mission purposes) and recognized that significant information may often be gained from relatively shallow drilling. The proposed drilling sites may be divided into five categories:

- Holes of opportunity from which scientific information may be gained without major expenditures for geophysical exploration or deepening the hole. Existing or planned holes may be scientifically productive at incremental cost in such locations as the Snake River Plain, the Reelfoot rift, and the Rio Grande rift.

- Holes of opportunity that may be deepened, for incremental cost, to respond to specific scientific questions. Examples are holes proposed to determine the subsurface boundaries of the Grenville province, in the Williston Basin, and in regions of the continental interior where basement rocks are undated.
- Sites where important problems have been identified by a substantial body of exploration data. Relatively expensive holes dedicated to scientific objectives can be drilled after further analysis of these data. An example is the San Marcos arch near Victoria, Texas. Information from a hole would answer fundamental questions about ancient continental margins and would complement offshore holes proposed in the report on Future Scientific Ocean Drilling (FUSOD).
- Sites where important scientific problems are to be found, but where substantial exploration is required before a decision on relatively expensive drilling is made. Many such sites—for example, in the Adirondack Mountains of New York, where the deepest exhumed crust known in North America may be exposed—can make major contributions to our understanding of crustal structure and evolution.
- Sites where substantial numbers of inexpensive shallow holes might be drilled to augment basic information for geological mapping, many of which would also be suitable for geophysical measurements. These holes would involve sites in many parts of the country where the basement rocks are hidden beneath glacial drift or have been subjected to deep weathering.

Thermal Regimes

Most of the heat that reaches the surface of the earth comes from the sun. This heat results in processes that erode and weather the continents, cause chemical fractionation of their materials, and tend to reduce them to sea level. Competing with the sun is the earth's internal heat, which drives convection in its interior, produces deformation of the surface, operates its own chemical "stills," and raises the continents above sea level. This competition has varied through time in a way that has not yet been firmly established or understood. Past products of the competition are the granitic rocks of the crust, mineral deposits, and petroleum basins on the margins of the continents. Present manifestations of the internal heat are earthquakes, volcanoes, and geothermal areas.

A major objective of earth science research is to understand the structure and dynamics of the various thermal regimes of the earth. Through observations we generate models of thermal systems. We test the predictive capability of these models of thermal systems with surface geochemical, geophysical, and geological techniques and refine the models so that they conform with the data. These data usually represent the surface results of the processes. We must drill in order to understand the dynamics of the processes, their behavior in three dimensions, and their past products.

The Panel on Thermal Regimes identified two major objectives. The first is to produce three-dimensional understanding of heat sources and products of thermally driven processes and to improve the boundary conditions of predictive models. The second is to remove the barriers to the understanding of high-heat-flow geothermal systems. Much research aimed at the first goal has been based on holes of opportunity; in view of the large federal drilling effort, it should be possible to continue and increase these investigations. The resulting data must be supplemented by information from boreholes drilled in locations specifically chosen to be responsive to scientific questions. In examining specific geothermal systems, dedicated holes will be required, but many of these will also be of value for the study of processes leading to the economic concentration of minerals.

Mineral Resources

After six millenia of mineral production we still do not understand how and why many kinds of mineral deposits form. In addition, as the more easily discovered surface and shallow mineral deposits become depleted, we must search ever deeper for their replacements. Investigations of physical and chemical characteristics of ore deposits and their geological settings at depth will derive great benefit from a systematic and focused program of continental drilling. This is particularly true for metallic mineral deposits.

The essential path to finding mineral deposits is to understand how the ore-forming processes have operated in the crust. Drilling and seismic studies are the only methods that can extend surface and shallow-depth studies to the depths necessary to involve related volumes of rock several hundred times greater than those of the deposits.

Many important mineral deposits are concentrations of valuable elements mobilized and transported with energy derived from hot magma (molten rock) driving reactions between aqueous fluids and rocks within the earth. Such centers of magma-geothermal activity may be sampled in depth by drilling in two types of situations:

- Currently active systems of interest in connection with fundamental principles regarding sources of geothermal energy. Examples: the Salton Sea area, southern California, The Geysers area, northern California, and the Yellowstone volcanic centers in Wyoming.
- Ancient mineralized hydrothermal systems that have yielded significant

ore deposits. Mining development of these deposits usually yields information in the immediate vicinity of ore. Scientific drilling in depth to explore several important mining districts will certainly provide significant new information bearing on their origins. High-priority districts include the following: Tonopah, Nevada; Tintic, Utah; Butte, Montana; and Santa Rita, New Mexico.

The nature of ancient sedimentary basins in which great deposits of iron ore were formed remains obscure. We lack sufficient understanding of the deeper strata in these basins, which is essential to interpreting conditions leading to formation of iron deposits. For example, a hole to probe deeply into the Animikie Basin of Minnesota could provide rewarding scientific data. Recent discoveries of high concentrations of heavy metals in brines of the deep sedimentary basin of the Mississippi Embayment offer an exciting opportunity for learning about processes that produce an important class of base metal deposits. The best examples of this class are the extremely rich copper deposits in Zambia and the lead and zinc deposits in Missouri. Similar deposits may be forming in the younger sedimentary basins of the Gulf Coast. A well-designed drilling and sampling program may provide an improved model for the formation of these deposits, which, in turn, could improve exploration strategies for a commodity essential to our high-technology economy.

Earthquakes

Three areas stand out in which a lack of fundamental knowledge concerning earthquakes has seriously impeded programs of generally accepted high national priority:

- The finding of suitable sites and techniques for isolating nuclear waste products
- The siting and earthquake-resistant design of nuclear power-generating facilities
- The siting and design of major dams.

Specific seismological research central to solving these problems includes seismicity distribution, earthquake prediction, seismic hazard assessment, and induced seismicity.

Most of the research employed thus far has been based on geophysical observations on the earth's surface, but resolution of critical problems requires that some observations be made in the third dimension, i.e., at depth. Drilling will be necessary for several reasons: (1) It provides an opportunity for study of physical properties of samples obtained from the neighborhood of the fault at depth. (2) It provides an opportunity for *in situ* measurements of stress and the variation of stress with time and location. (3) It provides an opportunity for measurements at depth, away from the effects of the surface, including weathered zones. (4) It provides an opportunity for making measurements from sources on the surface recorded by sensors at depth. (5) Certain measurements, such as those for fluid pore pressure, can be made only in boreholes. Relevant borehole observations include those of stress, temperature and heat flow, strain, fluid pressure and permeability, seismic parameters, material properties, and pore-fluid chemistry. Only by understanding the absolute-stress distribution in the earth's crust throughout the country will we be able to get a clear idea of the nature of earthquake-producing forces.

The U.S. Geological Survey is conducting a relatively comprehensive drilling program related to earthquake problems. Central to this program is the drilling of several deep and a number of shallower holes into the San Andreas fault in California. This program could be fruitfully expanded and extended to other parts of the country, including areas of apparently low seismicity, and could take account of the possibility of making relevant measurements in holes of opportunity that now exist or are planned for other purposes.

There is a need for an appropriate communication system to ensure rapid and thorough dissemination of information on drilling plans and available boreholes to interested scientists. Some of these holes need to be maintained, as it is important to measure not only absolute stress in holes throughout the United States but also stress variations with time. A major instrument-development program is required to develop more satisfactory means of measuring stress and its variation with time.

Program Costs and Cost Distribution

An analysis of costs for a scientific drilling program in the continental United States is not within the scope of this report. This analysis will depend on results of assessment of priorities for various classes of experiments, analyses of opportunities for various scientific combinations of experiments and their integration with mission-oriented drilling, establishment of the cost surcharges for adding science-related increments to various types of federal drilling programs, and projection of growth rates for both current drilling programs and science-related drilling programs.

For reference, the following are the drilling costs for various types of projects. A 100-m hole drilled for a heat flow measurement may cost $1000, whereas a 3-km hole in crystalline basement rocks can be a $2 million project. An important geophysical experiment added to a 3-km geothermal hole may be carried out for $50,000 (to cover an additional delay of one week and for operational requirements). Drilling experiments and testing technological capabilities at depths of 7–10 km may involve total costs in the range of $5 million to $30 million. Many problems and associated experiments discussed in this report and its appendixes lend themselves to cooperative drilling programs. With the proposed mechanisms for information and communication, many of these experiments may be completed as collaborative or coordinated expense-shared projects at a small fraction of the cost of an isolated dedicated drilling effort.

In establishing the initial phases of the Continental Scientific Drilling Program, an effective approach will be to provide funds for planning, organization, and appropriate opportunities on a progressive basis, compatible with the growing scale of federal drilling effort. As the scientific return is tested and opportunities for collaboration are identified, more specific budget priorities can be established.

The cost analysis should include costs for the advance science-related preparation for drilling, science-related activities during drilling and postdrilling analyses, and interpretation of results. Advance planning includes surface mapping, geophysical, geological, and geochemical investigations, and development of instrumentation, techniques, and procedures. Site selection for holes dedicated to scientific objectives must involve integration and interpretation of scientific data from several sources: (1) geological and geochemical investigations of the surface and inferences regarding subsurface conditions; (2) geophysical surveys that yield data on subsurface conditions—seismic, heat flow, electrical, magnetic, and gravity; (3) in certain cases, shallow drill holes; (4) theoretical modeling. Postdrilling activities frequently may require extended sample and data analyses. The scientific experiments associated with a dedicated scientific drilling project may represent one third to two thirds of the total cost of the project.

Funding for Science and Scientific Drilling Activities

A variety of funded scientific programs could begin to take advantage of some existing mission-directed federal drilling programs. As the magnitude of the scientific activity surcharges grows, their impact on budgets will be severe. It is therefore necessary to develop additional science funds for a significant level of collaborative activity, as well as to prepare for specific drilling experiments organized by the scientific community. Federal agencies whose mission effectiveness will be enhanced by science augmentation should consider initiating granting programs in support of the science phase. Granting agencies with earth science components should take into account the resources that may be required by basic science experiments less relevant to immediate mission-oriented agency needs and for the associated predrilling and postdrilling activities.

Drilling, Logging, and Instrumentation: Technological Capabilities and Challenges

The scientific experiments defined as especially valuable by the four scientific panels require a common set of capabilities for both drilling and downhole measurements. At present, holes can be and have been drilled and logged to 10-km depths with temperatures as high as 200°C. This level of technology can satisfy a large portion of the investigations suggested. Drilling at temperatures between 300°C and 350°C is considered a feasible extension of current technology and is currently being conducted on a limited experimental basis to establish the next level of capability. The drilling industry and the Department of Energy are conducting such development programs in support of the geothermal energy program. When this capability is available, the range of scientific drilling experiments will be extended into several important problem areas.

A further challenge to technological capability is already apparent in some of the proposed experiments, which include requirements for the following:

- A drilling system capable of operating to 500°C
- Logging techniques and instrumentation capable of making measurements to 500°C
- Improved techniques to determine the *in situ* state of stress in the crust; a noninvasive, repeatable approach is preferable
- Improved fluid- and solid-sampling techniques to obtain samples from deep within the earth.

To achieve these objectives, research is needed in the following areas: identification and/or development of materials suitable for high-temperature and high-reactivity environments; long-term fluid-flow measurement techniques; improved slim-hole drilling and continuous coring capability; and improved *in situ* neutron-activation analysis.

Overcoming these technological barriers to scientific exploration of the deep crust is a major challenge.

Implementation and Organization

The first step in creating a Continental Scientific Drilling Program is to determine what must be done to achieve the scientific objectives. This question has been treated by the panel reports that examined the scientific problems that may be resolved by drilling. They have identified problems, suggested sites, provided an indication of their relative importance, and suggested the scale of the costs. The need for a coordinated program is illustrated in that drilling sites chosen by the Panel on Thermal Regimes in the Salton Sea area would also be appropriate for critical studies on the origin of mineral deposits. Similarly, locations chosen by the Panel on Basement Structures would be appropriate for studies (suggested by the Panel on Mineral Resources) on metal-rich interstitial brines and sedimentary ore deposits. Scientific information from the programs of all of the panels will contribute to establishing better criteria for site selection and management of radioactive waste. Not only is it in the public interest to maximize the scientific return from the large public investment in drilling, but clearly an integrated effort is more valuable than the simple sum of its parts.

Given the need for coordination and sound scientific objectives, how shall we proceed? Who are the relevant constituencies, what are their needs, what can they contribute? The Panel on Implementation and Organization has proposed that a Board on Continental Drilling for Scientific Purposes be established as the central focus for the Continental Scientific Drilling Program and that the federal agencies involved in drilling establish internal offices to coordinate scientific activities within each, as well as an interagency coordinating body. The functions and responsibilities of the proposed board are detailed in Appendix F. The panel recognized that the growth of a significant Continental Scientific Drilling Program will inevitably lead to major investments of effort in developing both scientific advice and priorities and in instituting extensive operational functions to achieve maximum effectiveness of the program.

In the current stage of development of a Continental Scientific Drilling Program, a single board could serve as the mechanism for providing both scientific and operational guidance. But the anticipated magnitude of the total drilling program, with a mission-directed science orientation, could well lead to a more complex management structure to coordinate the interests, needs, and contributions of the academic, industrial, and governmental constituencies.

A continuing national scientific drilling program will require mechanisms to deal with two separate levels of analysis, advice, and operation. Two distinct but compatible committees might be appropriate: a high-level Scientific Drilling Advisory Committee (SDAC) to be concerned with science analysis, science program priorities, national drilling program objectives, and science program support; and a National Drilling Operations Committee (NDOC) to deal with information exchange, data and sample management, and, when desirable, arranging drilling operations for nonagency science investigations.

The SDAC could provide the following services:

- Identify and/or evaluate high-quality science programs involving drilling
- Identify and help establish sources of support for such science programs, where possible, by recommending them to appropriate mission-oriented agencies or to appropriate science-funding programs
- Identify opportunities for augmenting mission-oriented drilling programs with compatible science objectives
- On request, provide scientific and programmatic advice to government agencies
- Make recommendations on the general design, quality improvement, and integration of scientific drilling programs in the national earth science research effort
- Provide policy guidance to NDOC

SDAC committee membership would be based on scientific competence regardless of affiliation; familiarity with drilling as an investigative tool would be an important but not an essential qualification. Both mission-oriented and science-funding agencies would be invited to establish liaison representatives.

The NDOC would be composed of members from agencies with important drilling programs, the active scientific community, drilling-technology and instrumentation specialists, appropriate scientific and professional societies, industry, and state geological representatives. *Active participation or familiarity with current drilling programs for science should be a major qualification for membership in NDOC.* Nonfederal representation should be selected with the approval of SDAC.

NDOC would have the following primary but not exclusive functions:

- It would survey, receive, and disseminate comprehensive information to the scientific community, the public, and the government on active and planned drilling programs of various federal agencies and on opportunities for research through drilling.

- It would establish standards and arrange appropriate coordination of repository and catalog systems for cores, samples, and logs, with special attention to those of exceptional scientific value.
- It would establish a data-management system.
- It would assist interagency and intra-agency sharing of drilling expertise and opportunity.
- It would assist the scientific community in implementing research-oriented drilling programs recommended by SDAC by arranging integration into mission-directed drilling programs or by arranging independent drilling programs for scientific purposes only.
- It would provide a communications mechanism regarding drilling for commercial purposes to alert the scientific community to potential downhole research opportunities and the possibilities for extending the holes for scientific purposes.
- It would make recommendations to SDAC on policy matters involving all operational aspects of its program.
- It would periodically report to SDAC on its activities, its information base, and its experience with the state of drilling technology.

Both SDAC and NDOC could use auxiliary or *ad hoc* panels to help implement their functions. An adequate permanent staff and staff director could be shared by both committees.

The sustaining support for the two committees could be requested from various federal agencies with extensive drilling programs, on the basis that these committee functions would be cost-effective in their behalf. In addition, each major agency with drilling programs would be encouraged to establish internally a central information office for agency drilling activities and information.

Summary and Recommendations

The dynamics, structure, evolution, and genesis of the continents offer a major scientific challenge. This challenge has led international and domestic scientific bodies to plan programs for the 1980's in which crustal evolution and dynamics play a central role.

At the same time, society faces many problems that require scientific information about the continental crust for solution. Important among such problems are those relating to supplies of energy, water, and mineral resources; environment; isolation of hazardous wastes; and siting of dams, power plants and other structures for minimum risk from the effects of geological hazards.

A workshop held in 1974 led to a report setting forth objectives of a program of continental drilling in three broad scientific areas and called for development of a national program of drilling to achieve those objectives. In the intervening years, federal agencies have been responding to societal concerns with specific mission-oriented projects that include major drilling efforts that now call for expenditures of approximately $500 million per year. Thus elements of a national program of continental drilling now exist. *Clearly, it is in the public interest to maximize the scientific return from this very large investment of public funds.*

A Workshop on Continental Drilling for Scientific Purposes, which was held in July 1978 and which led to the present report, addressed the questions of how to maximize the scientific value of current and planned efforts of federal agencies and industry and how to supplement these efforts with holes drilled solely for scientific purposes. At the workshop, four panels addressed the scientific and associated societal problems. Their conclusions, appearing in detail in the appendixes, are briefly highlighted in this report. The subjects treated by those panels are as follows:

1. Basement structures and deep continental basins—broad and specific questions related to understanding the earth's continental crust.
2. Thermal regimes—basic understanding of geothermal systems.
3. Mineral resources—basic understanding of ore-forming processes.
4. Earthquakes—basic understanding of earthquake and faulting mechanisms.

The panels identified the main problems in each area whose solution depends on, or could be significantly accelerated by, information obtained from drill holes. They further addressed the use of holes of opportunity (holes drilled for other specific mission purposes), as well as holes dedicated to a specific scientific objective. Interrelations among the objectives of various panels are noted in this report and in the panel reports.

A fifth panel considered needs for technological developments. One principal immediate need is for development of drilling and logging equipment capable of operating at much higher temperatures (up to 500°C) than present capabilities (about 250°-350°C).

The importance of a communications and coordinating mechanism to maximize the scientific results of drilling projects—new, under way or planned, and proposed additional efforts—was reflected in the reports of all panels, especially the Panel on Implementation and Organization. To this end, a Continental Scientific Drilling Program is outlined, including two advisory and guiding committees, one concerned with scientific objectives, the other with operations (especially communications and coordination).

A scientific drilling program has three elements: *predrilling*—thorough exploration and problem definition; *drilling*—sample collection and downhole measurements; and *postdrilling*—studies and analyses. Program costs for scientific drilling projects are not addressed in detail, but it is emphasized that the associated scientific studies, including predrilling exploration, may constitute from one third to two thirds of the total cost of the project.

Principal Recommendation

The convergence of current basic scientific interests with the rapidly growing efforts of federal agencies to acquire, by drilling, the data needed for the missions that our society has given them; the need for an involved and well-informed scientific community to supply advice and interpretations; the need to maximize the effectiveness of research efforts based on an expensive tool; and the need for technological innovations to achieve program objectives—all these call for a mechanism to provide a central focus for continental scientific drilling and for communication and cooperation.

The U.S. Geodynamics Committee makes the following principal recommendation, that *a Continental Scientific Drilling Program be initiated to achieve expanded knowledge and understanding of the uppermost part of the crust of the earth in the United States for the objectives outlined in this report. This program would provide a central focus for the scientific aspects of federal drilling activities and a mechanism for communication and cooperation with academic, industrial, and state scientific constituencies.*

Bibliography*

Proposed United States Program for the Upper Mantle Project, Panel on Solid Earth Problems, Geophysics Research Board, National Academy of Sciences, 36 pp., Nov. 1962.
Solid Earth Geophysics: Survey and Outlook, Panel on Solid Earth Problems,

*Also see the complete list of cited references, page 184.

Geophysics Research Board, National Academy of Sciences, Publ. 1231, 198 pp., 1964.

"Deep Drilling on Land for Scientific Purposes," *Eos, Trans. Am. Geophys. Union, 47* (2): 373–378, 1966 (report of an *ad hoc* Committee on Deep Drilling for Scientific Purposes, a subcommittee of the U.S. Upper Mantle Committee).

Drilling for Scientific Purposes, report of a symposium in Ottawa, Canada, Sept. 2–3, 1965, Pap. 66-13, Geological Survey of Canada, Ottawa, 264 pp., 1966.

U.S. Program for the Geodynamics Project: Scope and Objectives, U.S. Geodynamics Committee, Geophysics Research Board, National Academy of Sciences, 235 pp., 1973.

Continental Drilling, E. M. Shoemaker, ed. Report of the Workshop on Continental Drilling, Ghost Ranch, Abiquiu, N. Mex., June 10–13, 1974, published by the Carnegie Institution of Washington, 56 pp., 1975.

Continental Drilling, Recommendations of the Panel on Continental Drilling of the FCCSET Committee on Solid Earth Sciences, 19 pp. plus attachments, April 22, 1977.

Report of the Workshop on Magma/Hydrothermal Drilling and Instrumentation, S. G. Varnado and J. L. Colp, eds. Report of a Workshop held in Albuquerque, New Mexico, May 31–June 2, 1978, Sandia Laboratories Publ. 78-1365C, 68 pp., July 1978.

Appendix A
Basement Structures and
Deep Continental Basins

A BASEMENT STRUCTURES AND DEEP CONTINENTAL BASINS

I. Introduction

A. The Problem

Less is known about the nature of the deep continental crust and how it developed than about the oceanic crust. Yet almost our entire sustenance and well-being are directly tied to this portion of the earth's crust. Our knowledge of the continental basement (as much as 60 km thick) is critically limited by lack of exposure. To extend our knowledge to the third dimension, geological, geophysical, and geochemical studies are needed. Although geophysics is the means by which much information has been obtained concerning this third dimension, drilling is the only means of directly testing the deep structures suggested by the other techniques.

Definition of the composition, distribution, and temporal and structural relations of the rocks that constitute the crystalline basement within the continental United States, as well as the processes involved in their formation, is a prime objective of American earth scientists. Such rocks, which may be as old as 3700 m. y., have undergone complex magmatism, metamorphism, and tectonism. Documentation of these events is required for a better understanding of the origin and evolution of the continental crust. Also needed is an understanding of the ways in which basement rocks relate to, and perhaps even control, features in younger rocks that influence the localization of ore deposits and seismicity.

Most of our knowledge of basement rocks has been gained from varied geological studies of rocks exposed at the earth's surface. However, our ability to answer or even to define important questions is hindered either by lack of exposure or by cover of the appropriate rocks by younger strata. Clearly then, a firsthand knowledge of the spatial relationships among rock types in the third dimension would greatly increase our capacity not only to understand the geological history of the crust but also to address environmental and resource problems more effectively. Such firsthand knowledge requires supplementation of surface observations with samples recovered from drill holes and by measurements and tests within the holes. Therefore the following recommendations are made:

24

- Drilling for scientific purposes should be an integral part of any national program for study of the continental crust.
- In addition to drilling purely for scientific purposes, ways are needed to capitalize on opportunities for extending or modifying holes drilled for other purposes. Such means require communication and coordination of all drilling efforts.
- Drilling must be considered as only one element of an integrated exploration program involving geological, geophysical, and geochemical approaches.
- Studies should be initiated immediately to analyze, integrate, and interpret samples and measurements from numerous exploratory boreholes drilled to the basement.

The crust has previously been interpreted as consisting of two to four universal layers. If the crust were indeed this simple, there would be little need for further detailed studies. The development of new knowledge and techniques makes it evident that the crust is far more heterogeneous and anisotropic. The complexity of the deep crust approaches that of basement outcrops at the surface; the only way to determine the structure of the deep crust is by means of deep continental drilling.

The development and growth of a continent is extremely complex. Of the 4 b.y. of geological history, the oceanic crust records only the last 250 m.y. The major steps in its evolution have been unraveled by extensive geophysical exploration and shallow sampling (with dredges and piston cores), with drilling programs such as the Deep Sea Drilling Project (DSDP) and the International Phase of Ocean Drilling (IPOD) as key elements to help identify the nature of widespread marker horizons and the underlying oceanic crustal types and ages. A wide variety of interactions among continents, island arcs and oceanic plates has been recognized. Continental crust is a result, at least in part, of the continuation of these processes through the enormous span of geological time. What today is a stable interior of a continent was at some time in its history part of the violent activity found along a plate boundary at a continental margin. Each present-day margin type has specific assemblages of rock types and magmatic activities, which in turn permit specific hydrocarbon and mineral associations. A basic goal of continental geological research is to unravel the evolution of continents—their growth and fragmentation. This framework can be invaluable to focus exploration for resources.

The purpose of this discussion is to address the following questions:

- What are the principal geological problems that require drilling?
- What can be learned about the structure, chemistry, and pressure-temperature-time relationships in unexplored regions of the continent?
- What is the present state of knowledge acquired from existing drill holes?

- What are the relations between drilling and other geological/geophysical/ geochemical studies conducted prior to and in conjunction with drilling?
- What kinds of studies can be performed in drill holes, in addition to studies on samples obtained by drilling?
- What specific areas of the defined problems are amenable to drill hole study?

B. Earlier Drilling Proposals and Some Results

The very terms used in this chapter title—basement structures and deep continental basins—imply geological features hidden from direct surface observation or unavailable to the shallow drill hole. Thus these features have been little known because of the paucity of deep drilling. Information hardly better than conjecture regarding these features has been available only in the form of extrapolation from surface or near-surface geological data and ambiguous geophysical data seldom controlled by the "ground truth" of direct observation. As a result, the geological mapping of basement structures and deep continental basins has remained at a rudimentary level, and little is known of the processes involved in their formation and subsequent modification. Understanding these processes is essential to the predictive goals of geological inquiry, which are so important in meeting the societal needs of today and of the future.

The paucity of information from deep drilling and the importance of rectifying this relative dearth have long been recognized by geoscientists. The "Basement Rock Map" of the United States (Bayley and Muehlberger, 1968) has several "no-data" areas, and the basement areas at 2 km or greater depth are generally known only by extrapolation. Numerous regional and national groups over the past decade and a half have proposed drilling programs that would augment our near-surface geological knowledge and complement the interpretation of surface geophysical data. These efforts culminated in the 1974 Ghost Ranch Workshop on Continental Drilling. Although the program outlined in the report of that workshop (Shoemaker, 1975) has not been implemented, the document resulted in a greater awareness of the significance of continental drilling in both the geoscience community and the funding agencies. As a result, for example, a consortium of university researchers joined together under a grant from the National Science Foundation to conduct a series of experiments in a deep (5.3 km) drill hole in the Michigan Basin, which was originally drilled by a group of petroleum companies. Results of these experiments and investigations on the cores obtained from the deep portions of the hole have added significantly to our knowledge of the Michigan Basin and have been useful in constraining the interpretations of surface geophysical data. But more important, the results of these studies are useful in evaluating models for the development of intracratonic basins on a global scale and permit enhanced interpretation of geophysical data by extrapolation from this "point source" of ground truth (Sleep and Sloss, 1978).

This is a particularly critical time to study basement structure and deep continental basins through drilling. The objectives of these studies become increasingly important as we search for deeper mineral and petroleum deposits. Also, it is becoming increasingly apparent that old concealed basement and deep structures may serve to focus, refract, or modify contemporary stresses. Basement structures extending deep into the crust may be especially effective. Resurgent tectonics along ancient structures is being recognized as a common phenomenon in the geodynamics of continental interiors. This was the theme championed by Flawn (1965) in his paper, "Basement: Not the Bottom but the Beginning," and continues to be emphasized by subsequent investigations. Furthermore, the concepts of plate tectonics have led to new models to explain the hitherto poorly known intraplate tectonism. Continental drilling projects that are critically sited by geological/geophysical investigations are needed to evaluate these models. Finally, new data from isolated mineral and petroleum drill holes, deep seismic reflection and refraction surveys, and in-progress compilations of the magnetic and gravity anomaly data of the United States are useful in defining the critical location of drill holes and in extrapolating information derived from these holes over large areas.

Virtually every piece of existing information on the structure of the continental basement—from surface mapping of outcrops to xenoliths to geomagnetic, seismic, and other geophysical data—tells us that the deep structure of the basement is complex. Folds, faults, unconformities, metamorphic boundaries, and intrusions are present in what today appears as bewildering variety.

But the data also tell us that the basement is not hopelessly complex. Most elements of the crust appear to have been affected by only a small number of major events (perhaps no more than three or four) in their history and not by the hundreds of episodes that would result in a snarl of information that might never be unraveled. It seems certain that at some time in the future our knowledge of the deep crust will be much more complete; our present state of knowledge of that topic will then appear rather primitive. The time of major advance may not be far off, for at present there exists a variety of methods for exploring the basement in much greater detail than has been possible in the past. To mount an appropriate effort is a feasible task of not unreasonable magnitude. It is therefore not too much to hope or even to forecast that, within the next few decades, prominent but currently unresolved structural features of the deep continental basement will become as familiar to continental geologists as the seamounts, ridges, plains, and deeps of the oceans have become to marine scientists in the decades following

World War II. We can visualize a time when subsurface basement intrusions, faults, metamorphic boundaries, and major folds are delimited, mapped, and even named. Unifying concepts comparable to plate tectonics will surely follow.

An era of intensive exploration of the continental basement is beginning. That such exploration will flourish in the near future seems beyond doubt, not solely because the continental basement is a major frontier of earth science, but because the increasing demands on our environment and the growing need for earth resources require a significantly improved adequate understanding of that portion of the earth upon which we depend for our livelihood.

This quest for understanding of the continents will involve a wide variety of earth science techniques—geological, geophysical, and geochemical. Drilling provides the ultimate geological truth with regard to identification of rocks and can provide critical information in studies of large-scale features. But because of the great cost involved in drilling, it must, in general, be integrated with other studies.

II. Major Deep Continental Structural Problems

A. The Precambrian Crust

The basic question about Precambrian basement crust concerns crustal growth—both vertically and horizontally. A related, but also fundamental, aspect involves crustal composition. Integration of chemical data on the continental and oceanic crusts yields the mean composition of material that was derived from the mantle to form the crust. Crustal composition must certainly vary vertically and laterally; these variations also change in position and time, recording the history of crustal growth. Drilling is the only sure means for determining composition in a vertical section.

How did the continents grow horizontally? They probably grew in later geological time by accretion of island arcs and collision of subcontinental plates, but the role of plate tectonics in the early Precambrian is much more difficult to assess. What is the significance of Precambrian greenstone belts? Do they mark suture zones? Are greenstone belts similar in composition to modern island arcs? Drilling is needed in each feature to resolve these questions. Do Precambrian crustal discontinuities (such as shear zones and petrographic boundaries) mark plate boundaries? What happened to the possible ophiolites that are commonly missing along these boundaries? Continental collision and subduction should produce dramatic structures such as abundant faulting, crustal wedges, strong lateral compositional changes, and, possibly, vertical differentiation through partial melting. A search for such structures offers numerous potential targets for deep drilling along possible suture zones.

Equally critical is the question of vertical growth of continents. Was crustal thickness determined during a relatively early consolidation during the Archean or Proterozoic? Or is material added to or subtracted from the continents later by underplating or subcrustal erosion? Is material transferred to or from the crust to compensate for vertical movements? The question of vertical movements is thus intimately related to crustal transfer. Is a "memory" incorporated in certain orogenic belts, which causes much later uplift? The nature of the deep crust beneath a deeply eroded basement terrane such as the Adirondacks presents a similar problem. These questions can be answered only by deep drilling and subsequent geochemical (particularly isotope) studies of cored material.

Estimates of the abundance and movement of water in the crust are both important and controversial. Water, even in small amounts, is highly effective for all material transport and especially for ore deposition. The abundance of water in the deep crust is controversial, because petrological considerations indicate the deep crust is dry, whereas geophysical data suggest that it is wet. Also, water concentration in the deep crust may be greater beneath former zones of island arcs. Drilling is needed to measure water concentration and flow in the deep crust.

What is the nature and significance of crustal discontinuities? Are there major crustal discontinuities—i.e., is there a "Conrad discontinuity?" Seismic refraction studies first resulted in the definition of the Conrad discontinuity. Some subsequent refraction studies have reported a Conrad, some have found none, and others have reported three to five crustal "layers." Recent seismic reflection surveys have found reflectors all through the crust; these reflectors vary greatly laterally, so the presence of one or more universal crustal discontinuities seems doubtful. Deep drilling may resolve this and questions of deep crustal heterogeneity and whether a mafic zone forms the lower crust. As more deep crustal reflection profiling is completed, some reflectors will probably exhibit great lateral continuity, as is suggested in the Hardeman County (Texas) study (Oliver et al., 1976). Such reflectors would be a logical target for deep drilling.

Any deep drilling should be preceded by thorough geophysical studies. Practically speaking, this means that Consortium for Continental Reflection Profiling (COCORP) deep seismic reflection profiling is desirable across any deep drilling site because of the unequaled high resolution of the reflection method as compared with other geophysical techniques.

Igneous intrusions are geological settings that may be profitably studied at present by drilling to modest depths (3–10 km). Are intrusions floored? If so, what is the nature of these floors? How are they emplaced? What has happened to the mass of country rock from the space now occupied by the intrusion? For a crust-derived intrusion, what is the nature of the source area? All these questions may be answered by drilling to depths currently achievable. In addition, the roof zones of felsic intrusions and the floors of mafic intrusions tend to be loci of concentration of metals, so that drilling of intrusions will lead to a better understanding of ore-forming processes.

B. Causes of Continental Tectonics

The physical processes accountable for continental tectonics are poorly understood as compared with those for oceanic tectonics. For purposes of discussion, it is useful to divide the continental crust into basement and supracrustal rocks. The basement consists of intrusive and high-grade metamorphic rocks that cannot be mapped as easily as supracrustal units because of their structural complexity or their igneous origin. Supracrustal rocks may be extensively intruded and metamorphosed, but may still be studied as layered units. Old supracrustal sequences are commonly overlain by subsequent ones.

1. Cratonic Basins and Arches

Cratonic basins including the Michigan, Illinois, and Williston basins lie within the interior of the continent and contain a few kilometers of nearly flat-lying sedimentary rocks. Areas of relative uplift surrounding the basins are termed arches. Extensive petroleum exploration has defined the general structure of cratonic sediments throughout the United States. Subsidence during the deposition of basin sediments rather than later folding is well documented. Hypotheses for the origin of these basins include the following:

a. Isostatically compensated sinking because of loading by sediments. This commonly accepted explanation does not account for the initial relief required to produce the observed accumulation of sediments. Even on the Atlantic Coast, where loading of oceanic crust may occur, a driving force in addition to sedimentary loading is necessary to produce the observed subsidence. Loading by sediments only modifies some other driving force.

b. Thermal contraction following a heating event. This hypothesis has received extensive consideration because the history of subsidence rate is predicted to be a decaying exponential and because thermally driven subsidence occurs at midocean ridges and probably at "hot spot" swells (e.g., near Hawaii). The Atlantic margin basin is related to continental breakup and spreading at the Mid-Atlantic Ridge. In contrast, heating before subsidence in interior basins is not so evident. Studies of drilled sediments are useful to analyze subsidence that accompanies sedimentation. Borehole gravimetry and well logging are useful in determining sediment density, which permits computing subsidence because of sediment loads. In addition, igneous rocks or metamorphism related to the heating event may be detected.

c. Crustal loading by igneous intrusions. This may occur as a separate mechanism during a thermal heating event. Rapid subsidence following the loading is expected. Drilling might detect this igneous activity.

d. Subcrustal erosion. Heating of the lower crust may be accompanied by return of crustal material into the mantle. This is a possible mechanism for thinning the crust prior to basin subsidence. The lack of subaerial erosion prior to subsidence is evidence that a subcrustal process thinned or loaded the continental crust. Various forms of this hypothesis are favored by Russian and European geologists.

e. Crustal necking. This is well established on Atlantic-type margins, particularly the Biscay coast of France. For interior basins, a space problem would result from this mechanism. Deformation in the substratum of basins may be detectable by seismic methods that are calibrated by drilling.

f. Phase changes and melt migration in the asthenosphere. The depths at which these processes occur are unreachable by foreseeable drilling methods.

Other than reactivation of older faults, the extent of control of basement structure on platformal basins and arches is unknown. This is partially because basement rocks are seldom penetrated in the deepest parts of basins. Drilling that is controlled by geophysical studies would extend our limited knowledge.

Arches are difficult to study because of the loss of sedimentary record. Shallow wells to determine uplift history of arches and to deduce "eustatic" changes in sea level may be warranted for some problems.

2. Folded Mountain Belts

Folded sedimentary sequences from Archean to Recent age occur in various parts of the United States. Although some are attributed to continental collision of subduction zones, many are not so easily explained by this concept. The growth of individual folds and the mechanics of thrusts are understood to a moderate extent, but the driving forces are not. Problems addressable by deep drilling include the following:

• Are there ensialic fold belts, especially in the Precambrian, that are unrelated to continental margin tectonics?
• What is the structure of the root zone of thrusts and nappes?
• To what extent is the basement involved in thrust tectonics? Where do detachment surfaces occur?
• What are the geometry and origin of mantled gneiss domes, and are they autochthonous?

- How does deformation of fold belts by igneous intrusion occur?
- How are ophiolitic and Franciscan-type complexes emplaced?

Because of the great relief and excellent exposures of all but the oldest fold belts (and those buried by later basins), the supracrustal deformation is moderately well understood in the foreland. Deep drilling is needed to define decollement surfaces and basement involvement, if any, but particularly hinterland relations. Oriented cores will be needed in most instances. Extensive geophysical work will be needed to extend the results by extrapolation away from drill holes.

3. Igneous Intrusives

Except for their bases, many igneous bodies are well exposed by erosion and thus can be studied without drilling. A main question in geology for over a hundred years has been: What is beneath the largest igneous bodies (batholiths and large layered intrusions)? Here, drilling appears capable of yielding useful information, but only with adequate geophysical surveys preceding site selection. In addition, shallow holes to obtain samples for age determination, geochemistry, and paleomagnetism appear warranted. Drilling near active igneous bodies is treated in Appendix B.

4. Rifts and Aulacogens

Rifts are defined simply as elongate depressions overlying places where the entire thickness of the lithosphere has ruptured in extension. Rifts are the commonest of the major lithospheric fault structures because the strength of the lithosphere is least under extension. The processes that lead to the extension and subsidence, however, are complex and often poorly understood. Proposed mechanisms include differential temperature changes in the lithosphere, phase changes, isostatic response to partial melting, subcrustal erosion, surface erosion following thermal uplift, injection and stoping of dense material, and collapse of linear magma chambers, alone or in various combinations.

Active intracontinental rifts and rifts with ages extending back as far as the early Proterozoic occur in North America (Figure A-1). Study of them should help identify the causative mechanisms, to show how the continental lithosphere ruptures in direct extension or oblique extension and how, once ruptured, the lithosphere continues to develop in places where oceans fail to form. Relating these numerous rifts to the various stages of the Wilson cycle—oceanic evolution from youth (continental rupture) to old age (continental collision)—appears feasible, and analysis in these terms seems a potent way of

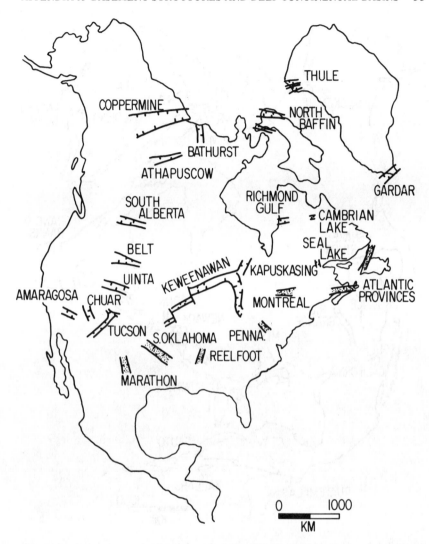

FIGURE A-1   Major Proterozoic and Paleozoic rifts of North America (Burke *et al.*, 1979).

applying the recent and revolutionary understanding of plate tectonics to the tectonics of the continents.

Among the numerous types of recognized rifts are those related to Atlantic-type margins, intracontinental rifts (e.g., Rio Grande, Snake River Plain) related to the Basin and Range province (Figure A-2), old rifts apparently exhibiting recent reactivation and seismic activity (e.g., Reelfoot), rifts

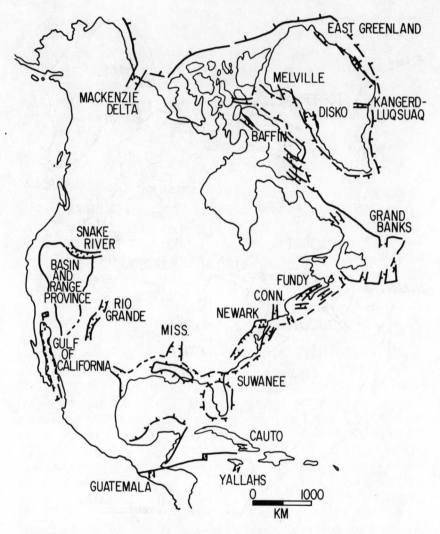

FIGURE A-2   Major Mesozoic and Cenozoic rifts of North America; most are associated with Atlantic-type ocean margins (Burke *et al.*, 1979).

represented by major gravity anomalies (e.g., Keweenawan), and putative aulacogens such as the Anadarko Basin.

III. Information Available from Continental Drilling

An immense variety of information on crustal structure and history of the earth's crust is potentially available from boreholes. Some of these data have been collected from wells drilled for oil; drilling for water and mineral exploration has also contributed.

Several types of material can be recovered from a borehole. Samples of the rock penetrated may be obtained in the form of cuttings, which are circulated to the top of the hole by the flow of drilling mud. Where higher-quality samples are needed, cores are taken. Such core material is preserved in full or as slabbed sections, although the economics of storage often dictates that only core chips (½-in. ($\approx$ 1 cm) slices taken at 12-in. ($\approx$ 30 cm) intervals) are preserved.

Test holes may be logged by a variety of instruments sent down the borehole on a wire line. In oil and gas wells these wire-line logs usually include electrical measurements (spontaneous potential, electrical conductivity), sonic measurements (seismic velocity), density (by a variety of techniques), gamma ray and neutron reflectivities, and dipmeter profiles. As new instruments and techniques are developed, these can be used to expand our knowledge of the rocks through which boreholes penetrate. For example, radioactivity may be measured by radiation detectors, and stress orientations may be determined by hydrofracturing.

Of the different types of information that may be obtained from boreholes, some are relevant to all holes and others are limited to specific types of rock penetrated (Table A-1). Lithology—determined from cuttings, cores, and (to a limited extent) wire-line logs—yields data on tectonics, structure (faults, contacts, uncomformities), and rock chemistry. Physical properties, obtained from wire-line logs, include density (for calibrating gravity models), seismic velocity (for interpreting seismic reflection profiles), magnetic susceptibility (for magnetic modeling), porosity, and permeability. Radioactivity allows determination of uranium, thorium, and potassium concentrations, and hence the volumetric heat generation of the rock. Heat flow is determined from the temperature gradient, measured after the thermal effects of drilling have subsided. Stress orientation, and sometimes magnitude, are obtained from hydrofracturing.

Microfossils obtained from cuttings or cores of sedimentary rock yield the geological age of the rock, while petrological studies give the rock's paleoenvironment, which, in turn—and in conjunction with seismic data—yields data on subsidence rates or crustal stability, dip of the strata, and diagenesis, which may lead to paleo-heat-flow estimates. Metamorphic rock cores yield information about metamorphic grade, hence pressure and temperature environments, and the radiometric age of metamorphism. Igneous rocks yield petrological information about rock genesis and radiometric ages of crystallization. Oriented cores of many rocks yield paleomagnetic estimates of the latitude and orientation of the rock when its temperature dropped below the Curie point.

A geologist interested in obtaining data from boreholes that have already

TABLE A-1 Types of Information Obtainable From Boreholes

	Geological Setting*		
	Sedimentary	Metamorphic	Igneous
Age	Paleontological age, deposition-erosion; 1, 2	Deformation age, radiometric; 1	Cooling phase, intrusive-extrusive event; 1, 2
Lithology	Clastic, carbonate, evaporite; 1, 2, 3	Metamorphic grade, tectonic setting; 1, 2	Genetic type, tectonic setting; 1, 2
Paleo-environment	Paleo-water-depth; 1, 2	Tectonic environment; 1, 2, 3	Environment of emplacement; 1, 2, 3
Seismic velocity	Calibration and interpretation of reflection data; 1, 3	Calibration and interpretation of reflection data; 1, 3	Calibration and interpretation of reflection data; 1, 2, 3
Density, porosity permeability, downhole gravity	Construction of gravity models; 1, 2, 3	Construction of gravity models; 1, 2, 3	Construction of gravity models; 1, 2, 3
Magnetic susceptibility	Construction of magnetic models; 1	Construction of magnetic models; 1	Construction of magnetic models; 1
Radioactivity	Heat generation and provenance; 1, 3	Heat generation; 1, 3	Heat generation; 1, 3
Alteration history	Diagenetic history; 1, 2	Postmetamorphic history; 1, 2	Postintrusive history; 1, 2
Thermal gradient and thermal conductivity	Heat flow; 4	Heat flow and heat generation; 4	Heat flow; 4
Structure	Dip of beds, contortion-nature of contact; 1, 3	—	—
Paleomagnetism	Redbeds: orientation and paleolatitude; 4	—	Paleolatitude, orientation, age; 4
Resistivity	Lithology, fluid characteristics; 3	Mineral variation and fluid properties	Mineral variation and fluid properties
Stress-strain measurements	Current tectonic state; 3	Current tectonic state; 3	Current tectonic state; 3
Equipment emplaced in hole	Seismic, temperature, magnetics, stress measurements	Seismic, temperature, magnetics, stress measurements	Seismic, temperature, magnetics, stress measurements

*Numbers refer to source of information: 1, cores; 2, cuttings; 3, logs; and 4, special measurements.

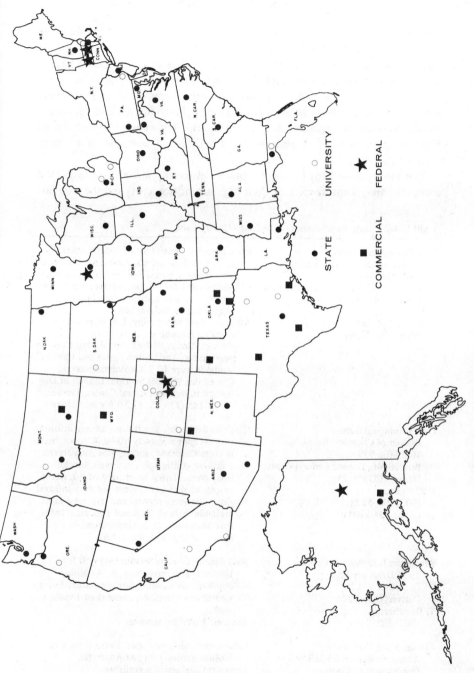

FIGURE A-3   Sample and core repositories as of 1976.

37

been drilled can employ a number of resources. County lease maps show the location of all wells that have been drilled in a given area, along with the name of the operator, the total depth, and the date of completion. Detailed information about the well, including cores, cuttings, and logs, is scattered among numerous state, federal, private industrial, commercial, and university repositories. Information about wells drilled by individual federal agencies is usually kept by the agency. Accompanying this report is a map of United States core repositories (Figure A-3) and a listing of borehole data repositories (Table A-2).

The information kept by these repositories usually includes cuttings (full cores are rarely kept in spite of their high acquisition cost) and the wire-line

TABLE A-2   Data Repositories on Drilling—Mainly Land

U. S. Geological Survey
Branch of Oil and Gas Resources
Mail Stop 940
Box 25046, Denver Federal Center
Denver, CO 80225
C. W. Spencer
(303) 234-3893
(FTS: 234-3893)

Core libraries—central list of federal, state, Canadian, and industrial repositories that hold well samples, cores, logs, and other data for wells drilled for oil, gas, mining, water, construction, waste disposal, etc., for the United States and Canada.
Services: Hard paper copy. List of available cores on request.
Reference: "Index of Well Sample and Core Repositories of the United States and Canada" available from U. S. Geological Survey, Open Files Services Section, Branch of Distribution, Denver Federal Center, Denver, CO 80225 ($29.25; $3.50 for microfiche).

U. S. Geological Survey
Branch of Chemical Resources
Mail Stop 939
Box 28046, Denver Federal Center
Denver, CO 80225
J. K. Putman
(303) 234-5121
(FTS: 234-5121)

Oil-shale drill cores as a file containing information on approximately 300 drill cores, mainly from Colorado, and representing almost 1 million stratigraphic intervals. Fischer-assay data relating to amount of oil, water, gas, spent shale, etc., given for each interval.
Services: Resource reports and magnetic tapes available through National Technical Information Service, U. S. Department of Commerce, Springfield, VA 22161.

Petroleum Information Corp.
1375 Delaware St.
P. O. Box 2612
Denver, CO 80201
T. Dougherty
(303) 825-2181

Well History Control System (WHCS) for all United States; inventory, by location, depth-status classification, operator, description of cores, etc., for more than 1 million wells.
Services: Full data services.

Florida State University
Antarctic Research Facility
Department of Geology
Tallahassee, FL 32306
D. Cassidy
(904) 644-2407

Information on cores from DVDP (Dry Valley Drilling Project) for the Antarctic.
Services: Core samples available.

Oregon State University
Department of Geology
Corvallis, OR 97331
P. S. Yeats
(503) 754-2484

Well logs, adjacent to active faults in California.
Transverse Ranges; cores of basement rocks,
Los Angeles Basin.
Services: By arrangement with institution.

University of California
Lawrence Livermore Laboratory
P. O. Box 899, L-224
Livermore, CA 94550
H. W. Howard
(415) 423-6480
(FTS: 532-6480)

Data bank of approximately 4,000 boreholes
from the Nevada Test Site and various
energy resource development and nuclear
excavation projects. Data include a catalog
of 20,000 geophysical logs containing geological description, explosion effects, and
hole description.
Services: Full data services available (listings and
plots with special search criteria).
Reference: USRL-78799, The LLL Nuclear
Test Effects and Geologic Data Bank, 1976.

California State College, Bakersfield
California Well Sample Repository
9001 Stockdale Highway
Bakersfield, CA 93309
J. R. Coash/J. Tucker
(805) 833-2324

Rock samples representative of the geologic
history, stratigraphy, rock properties, and
mineral resources of California.
Services: Collect, store, preserve, and make
available for public inspection rock samples
from 1900 oil and gas wells.
Reference: California Well Sample Repository
"Catalog of Well Samples," November 1977.

Louisiana State University
Department of Geology
Baton Rouge, LA 70803
D. Kupfer
(504) 388-3353

Information on rocks and minerals of the
Caribbean Sea and Gulf of Mexico; information on wells and stratigraphy for Louisiana. Geology literature for southeastern
United States.
Services: Further information available upon
request.

The information given in this table is a preliminary listing, taken from the *Directory of
U. S. Data Repositories Supporting the International Geodynamics Project* (World Data
Center A, 1978).

logs. For many scientific purposes, full cores are often necessary or desirable.
Certainly for exceedingly deep holes all core material should be kept.

The U.S. Geological Survey has recently proposed the creation of a core
repository that would retain full cores. It now stores well-log data for about
one million wells. However, it does not regularly receive data on wells drilled
for scientific or other purposes by other federal agencies. An important part
of an organized scientific drilling program would be to ensure that data are
collected and made available to the scientific community.

Drill data now available have not been fully exploited; much may still be
learned from the study of cores, cuttings, and other information now stored
in the various repositories noted above. Funds should be made available for
the study and synthesis of these data.

IV. Roles of Drilling

The principal scientific role of drilling, which is one of the most costly methods for investigating the earth's interior, is to solve problems. Commercial use of drilling by mining and petroleum companies to prove and measure mineral and energy resource reserves is well-known. Scientists have been able thus far to use drill holes to solve selected scientific problems only on a modest scale. As man's use of the earth, particularly the continents on which he dwells, outstrips his ability to cope with resulting problems, the need for new earth science information and concepts grows enormously. Drilling is only one avenue, but it is the most direct; it must, however, be used in concert with numerous other indirect scientific methods. Use of the other methods not only permits definition of problems and the selection of suitable drill sites, but also permits extrapolation and extension of the data from a drill hole to other areas and regions, thereby maximizing the benefits to accrue from a drill hole.

The uses of drilling are numerous. It is the only method for direct sampling of earth materials at great depth. Cores and drill-bit cuttings may be examined and analyzed in numerous physical and chemical ways so that their compositions and properties may be ascertained and proved. In addition, the fluids from various depths may be analyzed. Thus inferences regarding the materials, structure, and ages of the various levels in the earth's crust may be tested.

Drilling is used to define the structure of the earth and its materials. One drill hole may define the depth to the top of a particular rock layer, but only at that place. A second hole, by showing the depth of that same layer at another locality, may help to show the inclination of that layer. If available geophysical methods have been unable to provide adequate information, additional holes may be needed to define its configuration.

Drilling is also used to provide holes or windows to inaccessible environments. Despite the great ingenuity of laboratory scientists in devising and performing experiments at elevated pressures and temperatures, conclusions from these experiments are based on inferences regarding conditions and chemical systems within the earth's crust. Only direct measurements of true conditions at various depths will permit testing of the validity of the experimental data. A borehole permits the emplacement of numerous tools at selected depths to measure the many physical conditions (such as temperatures and pressures of both rocks and fluids, stresses, magnetic orientations, radioactivity, electrical fields, velocity, and conductivity) that bear on scientific, environmental, and resource problems.

The ultimate purposes of drilling are to define the materials, structures, conditions, and processes in the third dimension—depth—and to date these materials in order to determine the history of events or processes. However, before the third dimension is attacked at considerable cost, surface relations should be well described and defined so as to provide a basis for selecting critical problems to be tested at depth.

A. Predrilling Geology, Geophysics, and Geochemistry

The varied kinds of geological, geophysical, and geochemical observations that may be made at the earth's surface permit recognition and definition of geological problems that warrant attack to meet society's needs. Such investigations are mandatory to ensure maximum benefits of drilling and to provide adequate bases for the selection of appropriate drill sites.

Among the most useful of surface observations are those that relate to the composition, age, distribution, and configuration of earth materials at and just below the earth's surface. Such observations are normally collected systematically and coherently in the process of making a geological map. The geological map is not only an effective scientific tool but also a succinct form of communication, for it conveys to the map user the data, information, and concepts developed in the course of the surface geological investigation. Moreover, a well-made modern geological map permits projection (or extrapolation) to depth of the observed surface rock relations.

The buried, or subsurface, implications of geological mapping may be further tested and extended by geophysical surveys made on the earth's surface. Reflections of sound (seismic) waves from various rock layers at depth, and measurements of the gravity, magnetic, electric, stress, and thermal fields at various localities, are all clues to buried rock layers and structures, but all require various assumptions for interpretation. Nevertheless, the effectiveness of these methods has been proven in numerous instances, and they provide further means for defining both problems and sites that warrant further study by drilling.

Composition of earth materials is defined not only on the basis of physical and, at times, paleontological examination and description, but also in terms of chemical analysis. Chemical compositions and isotopic ratios reflect not only the conditions and times of rock formation; some elements and ratios bear inherited imprints that reflect the composition and age of the part of the crust or mantle from which they came. Thus geochemistry is an important part of numerous geological investigations. Moreover, geochemistry is essential to the studies of samples of the deeper crust and mantle brought up from below as fragments or as xenoliths in igneous rocks that, too, are clues to parts of the crust that warrant further testing by drill holes.

Integrated analyses by multidisciplinary teams of earth scientists are thus required to define geological problems warranting investigation and to select the sites most amenable to their solution. Such teams are likely to be aware of localities on the earth's surface where some important problems can be solved without resorting to costly drilling. Other problems may be amenable to solution at places within the reach of shallow drilling. Moreover, inexpensive shallow drilling is an effective means of improving the quality of geological, geophysical, and geochemical data for numerous parts of the nation, thereby expanding our knowledge of the regional geology. Nevertheless, some problems will require deep drilling.

The current status of required investigations across the United States is highly variable, depending upon the specific observation and the geographic location. Currently, only approximately 40 percent of the U.S. surface is mapped geologically on a regional basis. In the subsurface this percentage of map coverage decreases exponentially with depth. Basement depth, rock type, geochemistry, and isotopic age are mapped in only a rudimentary fashion and are available only sporadically from depths exceeding 2 km. Geophysical measurements that are particularly important in studying basement structures and deep continental basins cover a broad range of seismic, magnetic, gravity, electrical, and thermal measurements, but regional measurements are unavailable except on a localized basis.

Improvements are required in the current data base to complement the program of drilling basement structures and deep continental basins, and certain geophysical measurements are particularly needed because of their use in deep geological mapping. Specifically, seismic reflection studies complemented by seismic refraction profiles are needed to localize drilling targets within the deep continental basins and the basement. Such studies are expensive, but must precede the drilling. The seismic reflection method is particularly valuable in this respect. An example of the results of the COCORP study of a test profile in Hardeman County, Texas (Figure A-4) illustrates the ability of the method to show coherent mappable reflections from geological interfaces in the deep crust and upper mantle. If the coherent reflections at approximately 3 and 4 s travel time continue laterally for significant distances, they possibly may be traced to areas where they are exposed or are at relatively shallow drillable depths. If so, definition of their composition and structure by drilling or sampling would provide important information for understanding widespread parts of the deep crust. Additional seismic reflection studies are urgently required to complement the Continental Scientific Drilling Program. Aeromagnetic surveys are also extremely useful in mapping the depth to basement and the basement lithological zones, which in turn can be used in structural analysis of the basement rocks. These and other more specialized uses of aeromagnetic data make it necessary to improve the geomagnetic map coverage of the United States. Despite a large number of individual magnetic surveys over slightly more than half of the United States, only approximately 20 percent of the United States has been surveyed in a consistent manner. An improved magnetic anomaly data base covering the entire United States would greatly assist the several phases of the Continental

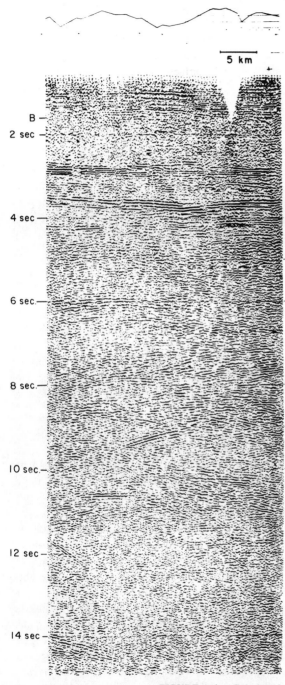

FIGURE A-4   Results of COCORP test profile in Hardeman County, Texas (Oliver *et al.*, 1976).

43

Scientific Drilling Program. Thus an ongoing seismic reflection study of the basement to determine primarily horizontal boundaries, and a magnetic surveying program to map near-vertical contacts within the basement rocks would complement each other and, supplemented by other geoscience surveys and measurements, would enhance the results of drilling basement structures.

B. Scientific Support During Drilling

Monitoring of drill holes as they are drilled is mandatory to assure appropriate testing and sampling. Initial logging (description) of cores and cuttings by geologists at the well site provides immediate examination of the validity of the geological model being tested. If the particular rock or structure that is the drilling objective is encountered at shallower depth than initially anticipated, drilling can be stopped. Decisions regarding appropriate levels to be cored are normally made at the drill site, based on knowledge of materials already encountered; decisions are similarly made regarding the need for oriented cores. Decisions regarding the depths at which formation fluids will be sampled are determined both before and during drilling. The drilling methods used require consistency with objectives; if geochemical analyses are a goal, contamination by inappropriate drilling muds must be avoided.

For some boreholes, concurrent laboratory support and analyses may be deemed advisable during the drilling. Moreover, some borehole geophysics can be scheduled at selected intervals during the drilling, to test for physical properties of critical horizons.

C. Postdrilling Geology, Geophysics, and Geochemistry

Data from any borehole may normally be extended to various distances from the drill site. If the drilling objective is to identify the composition and depth of a regionally extensive reflective or resistive layer in the crust, for example, then the borehole information is immediately applicable to a large region.

Thorough analyses of both materials drilled and tests within the hole are objectives of the drilling. Such analyses require time, but normally represent only a small fraction of the total costs. Further, unless such analyses are scheduled and actually performed, maximum benefits will not have been elicited from the drilling.

Various geophysical surveys of each borehole are advisable, not only as appropriate to immediate purposes of the drilling, but also for unanticipated later uses of the drill hole data. Thus careful attention is required for the selection of appropriate logging techniques having greatest possible application. Such logs provide the velocities of the rocks drilled (to reinterpret already completed seismic surveys) or resistivities and conductivities (for more accurate interpretation of electrical surveys).

Complete mineralogical, petrological, and physical descriptions of the materials drilled are mandatory, because they provide the basic record of the rocks sampled during drilling. Selected critical layers, rather than all materials, also require complete chemical and appropriate isotopic analyses.

Measurements made on a recovered core should include sonic velocity measurements (V_p and V_s), *in situ* pressure and temperature, electrical and thermal conductivity, porosity, permeability, strain characteristics, petrofabric and paleomagnetic studies (when oriented cores are available), and a thorough petrological and geochemical characterization of the core as a function of depth. These measurements will serve a twofold purpose. First, they will place constraints on the accuracy of surface and downhole measurements made before and during drilling; second, they will provide invaluable insight into the nature of that part of the crust that is otherwise unavailable for study.

A further application of such measurements is to test the suitability of the different levels of the crust thus sampled for purposes such as nuclear waste disposal or storage of nonrenewable resources such as gas and oil.

Initial analysis of a completed drill hole frequently reveals the need for further sampling of earth materials. Side-hole coring is a means of sampling rocks, and use of packers at various levels permits the sampling of fluids at appropriate levels.

Measurements of stresses in rocks within the hole may be made in various ways, but the currently preferred method is hydrofracturing. A final step, after drilling, may be the emplacement of tools in the hole to permit further recording and transmittal of data.

Synthesis of all data gathered during and after drilling is, of course, mandatory to evaluate their implications to the major geological problem defined for testing by drilling. Furthermore, the new information, when combined with the results of the initial studies, greatly extends the implications of completed geological mapping, geophysical surveys, and geochemical analyses. Evaluation and synthesis of all data may well result in the development of important, needed new concepts applicable to the needs of society.

D. Shallow Holes and Trenches

Much of the northern United States is covered with glacial deposits, which obscure the relationships of underlying rocks. Thin modern alluvial, marine, and fluvial deposits, as well as Cretaceous outliers, cover other large regions. Rock outcrops are often poorly exposed, and extensive weathering may extend deeper than 100 ft ($\approx$ 30 m).

Road cuts, as well as dug trenches and boreholes, have proved invaluable in routine geological fieldwork. The mining and stone-quarrying industries routinely use shallow holes to appraise deposits. Possible uses of scientific drilling or trenching include extending mapping into poorly exposed areas, obtaining stratigraphic sections of poorly consolidated rocks or rock units having low lateral continuity, and obtaining unweathered samples for geochemical, paleo-magnetic, or physical property studies. The costs of shallow holes and trenches are often small and comparable to other expenses in geological field-work for scientific purpost. An outline of current and possible uses follows.

1. Geological Contacts

Geological contacts are commonly zones of weakness that become poorly exposed valleys and swamps. Drilling of geological contacts would be highly useful in deducing age relations between large rock units, especially where there is little topographic relief for providing three-dimensional control. Major Precambrian-province boundaries in Wisconsin, Minnesota, and Michigan are possible sites for the drilling of geological contacts. These areas have been extensively mapped, and gravity and magnetic data are often available.

2. Carbonate Reefs

Silurian carbonates outcrop in the area surrounding the lower peninsula of Michigan. The generally flat topography and poor exposures in this region have hindered the understanding of these rocks. Reef buildups of limited spatial extent that occur throughout the region have been extensively quarried for stone and drilled for hydrocrabons at depth. The Indiana Geological Survey routinely drills below the floor of quarries to improve stratigraphic knowledge. It has now been determined that several stages of reef buildup and die-off occurred and that the reef formation was roughly contemporaneous with salt deposition in Michigan.

3. Metapelitic Rocks

Pelitic rocks in metamorphic terranes are easily eroded and inadequately exposed as compared with quartzite. Metamorphic fabric and mineralogical changes are best displayed in pelitic rocks, but the regional structure in pelitic sequences is thus often hard to deduce. Seismically active faults in such regions often follow regional foliations or older lithological contacts (e.g., near Mina, Nevada).

A few shallow holes, road cuts, or trenches are sufficient to determine the homoclinal component of bedding and foliation. Oriented cores are necessary for this purpose, unless several holes in different directions are drilled and the structure is consistent within the area of the holes.

4. Poorly Consolidated Sediments

Poorly consolidated sediments are seldom well exposed. Shallow trenches are used to study river and glacial deposits. Glacial deposits have been routinely drilled by private and state agencies for groundwater and engineering purposes. Semiconsolidated sediments such as Cretaceous outliers in Minnesota and Jurassic redbeds in Michigan could be studied easily with a few boreholes. Trenching of alluvial fault scarps is now routinely done in the western United States (Sieh, 1978).

5. Unweathered Samples

Remanent magnetism in some areas is adversely affected by weathering. Some shallow holes have been drilled to obtain unweathered sections through magnetic reversals in redbeds.

Rock-property and geochemical studies often require unweathered samples. In glaciated areas a hand-held drill may be sufficient to obtain samples below weathering. In the past, samples obtained for other purposes have been used for geochemical and mechanical studies.

In view of the above, it is recommended that a shallow-hole (less than 50 m) program be initiated to aid in the solution of special problems, including those related to geological mapping.

E. Ultradeep Drilling

Current technology permits drilling, logging, and instrumentation to depths of 10 km. Consideration should be given to drilling to depths of 15 km or more, for both practical and scientific reasons. Though inevitably costly, any such hole would be both an opportunity for the advancement of technology and for the long-term testing of instrumentation at high temperatures and pressures and an opportunity for long-term monitoring of heat flow, seismicity, and other deep crustal phenomena. For these purposes, ultradeep holes would require the considerable added expense of permanent casing.

The scientific purposes of such a hole would be primarily to serve as a standard reference section of the complexities of the basement and as a standard for calibrating numerous geophysical methods performed on the surface as well as in the hole.

V. Selected Targets for Drilling

Exploration of the continental basement is a huge task. Drilling into the basement almost anywhere will provide useful information of some sort. It would be wasteful and inefficient, however, to drill holes at locations that do not appear to be more promising for the production of scientific data than other sites. We must use available information to develop a plan for drilling that is designed to optimize our chances for obtaining the most significant results.

This section contains a list of selected drilling targets (Figure A-5) with brief descriptions. This list is representative, not exhaustive; it may not include some sites of comparable or greater merit. The sites are examples of substantial variety and hence may be used to illustrate the range of considerations necessary in developing a drilling program.

The drill is but one of many tools used in crustal exploration. When that tool should be used is an important consideration; rarely should drilling be used without preliminary exploration by one or more other methods, because it is expensive. It should be used only when required information cannot be obtained in less expensive ways and when the information sought appears likely to be of sufficient value to justify the expense.

Special situations may sometimes arise for add-on drilling and experiments where drilling for scientific purposes alone might not be justified. For example, a hole drilled to moderate depths for other purposes might—as a hole of opportunity—be extended beyond its normal target depth to probe a deeper horizon of scientific interest. Perhaps the location of the hole drilled for other purposes might not be critical and hence might be selected on the basis of the opportunity for a scientific add-on experiment.

Determination of the significance of a particular scientific problem is a matter of judgment and reevaluation in the light of new information. The value of drilling a particular hole might be solely scientific, or it might be both scientific and practical. All such considerations must be taken into account in evaluating a drilling plan, and continual reevaluation must occur before and during drilling. The potential for obtaining new knowledge of a site from further exploration is another factor that affects a decision on drilling.

The sites listed below, and discussed later, were classified with the above considerations in mind. (The numbers are keyed to the sites on Figure A-5.)

Sites ready for scientific drilling at present and without additional exploration—could be relatively inexpensive.
   Snake River Volcanic Province (1)
   Reelfoot Rift (2)
   Rio Grande Rift–Socorro Area (3)
Site ready for scientific drilling after further analysis of existing data to pinpoint location—requires new hole, relatively expensive.
   Crustal Structure of the Gulf Cost and Southeastern United States (4)

FIGURE A-5 Selected drilling targets. (Numbers correspond to sites discussed on pp. 48–80. Locations of proposed holes for Continental Interior Basement Mapping (7) and for the Geodynamics of Iceland are not shown.)

Sites where relatively inexpensive bottom hole extensions are desirable—if suitable add-on experiment opportunities arise.

Grenville Front (5)

Basement Rocks of Cratonic Sedimentary Basins (6)

Continental Interior Basement Mapping (7)

Sites where important scientific problems are known, but where substantial exploration is required before a decision on drilling is made—drilling will probably be relatively expensive.

Setting in the Adirondack Region (8)

Nature and Extent of Archean Rocks in Minnesota (9)

Nature of the Continental Crust of Southern California (10)

Third-Dimensional Relationships of Xenolith Occurrences (11)

Columbia Plateau—Nature of Volcanics and the Underlying Crust (12)

Thrust Fault Geometry and Its Relation to Basement (13)

Stable Interior Crystalline Terranes—Stress, Rheological, and Petrological States (14)

Precambrian Crustal Discontinuity in Southeastern Wyoming (15)

Character of the Crust in the Western Great Basin (16)

Tectonostratigraphic History of the Wichita Mountains Aulacogen (17)

Sites where substantial numbers of inexpensive shallow holes might be drilled to augment outcrop information for mapping or other purposes.

Many sites throughout the country, particularly in areas of glacial cover or deep basement weathering.

A. Snake River Volcanic Province (1)

The geometry and genesis of continental rift systems, such as the Snake River volcanic province and the Rio Grande graben, are so inadequately understood that they continue to be a source of controversy. Yet better understanding of these major tectonic features is necessary for improved assessment of earthquake risks, volcanic hazards, geothermal resources, the volumes of associated aquifers, and proliferating land use decisions.

The Snake River volcanic province is undergoing active exploration by both government and industry for potential geothermal energy. A moderately deep test, to be started by the Department of Energy in the fall of 1978 at the Idaho National Engineering Laboratory, offers the potential for capitalizing on that investment with add-on experiments. The hole depth is now planned at 3 km. Whether it will penetrate the entire sequence of the thick pile of volcanic rocks (both basalt and rhyolite) underlying the plain is uncertain. Yet a most crucial question is the composition, age, and structure of the rocks directly underlying the upper Cenozoic volcanics and sediments exposed on the plain. If they are not encountered in the planned borehole, deepening of that hole would be warranted to help discriminate among the several geometric and genetic models devised for rift systems.

Moreover, knowledge of older rocks underlying the Snake River Plain would permit far better interpretation of numerous geophysical surveys already completed and now in progress for the region. These include extensive seismic refraction profiles both along and across the axis of the rift system.

Data from such a hole would be of interest in connection with scientific and economic questions. A foremost question is: Do the folded thrust plates now producing oil in Utah and western Wyoming extend northward beneath the Snake River Plain into Idaho? Another is: Do the partly mineralized horsts and graben bounding the plain on both sides extend into the plain?

B. Reelfoot Rift (2)

The northeast-striking Reelfoot rift (Ervin and McGinnis, 1975; Hildenbrand *et al.*, 1977) at the head of the Mississippi Embayment has been interpreted on the basis of geophysical data and assigned a late Precambrian age. Subsequently, this area developed into the Paleozoic Reelfoot Basin and has been associated with rifting in the Mesozoic leading to the Mississippi Embayment aulacogen. The New Madrid earthquakes of 1811 and 1812 were among the strongest ever to occur in the United States. Contemporary seismic activity in the New Madrid area (Figure A-6) that is closely related to this rift may signal its reactivation. Understanding the geological history of this area is particularly important because of the relation to problems of contemporary seismicity. Further, this area is near the Mississippi Valley lead-zinc deposits; evidence of evaporite deposition in the basin and high-temperature diagenesis would aid in interpreting the origin and distribution of lead-zinc deposits.

Currently, this area is being subjected to intense study by the U.S. Geological Survey and the U.S. Nuclear Regulatory Commission–New Madrid Study Group. Investigations include intensive surface geological studies, gravity, magnetic, microseismicity, high-resolution seismic reflection, and crustal seismic studies. These studies will provide the control for the specific location of the drill hole, but tentatively the hole should be drilled in southeastern Missouri at the intersection of Reelfoot rift and Pascola arch. The hole should be targeted for a depth of approximately 5 km.

The major objective of the drill hole will be to evaluate the geophysical interpretation of the occurrence of a rift (graben) associated with the New Madrid seismic zone and the head of the Mississippi Embayment, to determine its age from the isotopic study of the igneous rocks and paleomagnetic investigations of both the sedimentary and the igneous rocks, and finally to investigate the metamorphic rank of these rocks as an aid in constructing models of the development of the rift and subsequent basins. The U.S. Geo-

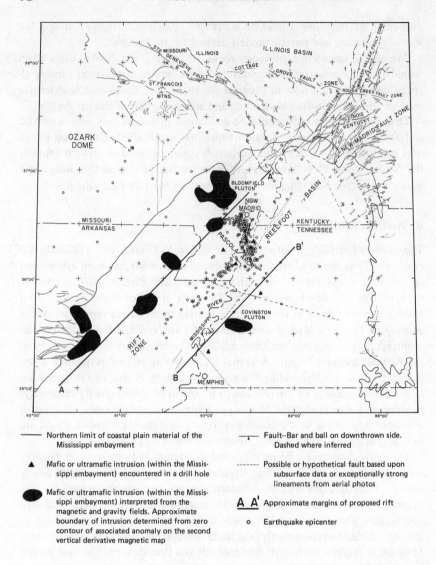

FIGURE A-6   Reelfoot rift (Hildenbrand *et al.*, 1977).

logical Survey has planned several drill holes within the rift that will penetrate
the Mesozoic rocks and approximately 0.3 km of the Paleozoic rocks. These
holes will be useful in specifying the proposed drill hole site and may permit
add-on experiments.

C. Rio Grande Rift–Socorro Area (3)

1. Geological and Geophysical Characteristics

The area near Socorro, New Mexico, is located within the Rio Grande rift, a major structure formed by east-west crustal extension beginning about 27 m.y. ago and continuing to the present (Chapin et al., 1978). North of Socorro the rift consists of a series of linked north-trending structural depressions arranged en echelon in a NNE direction into central Colorado. In southern Colorado and northern New Mexico the rift penetrates the southern Rocky Mountains. In central New Mexico the rift lies between the Colorado Plateau and the High Plains. South of Socorro, the rift merges in a complex and unknown way with the Basin and Range province. Geological evidence cited by Chapin and Seager (1975) suggests that rifting is occurring along an ancient line of weakness in the continental crust.

In the Socorro area an unusual geological characteristic of the rift is prominent intragraben horsts that formed relatively late in the history of the rift, 7 m.y. ago (Chapin et al., 1978). In the immediate vicinity of Socorro the rift is superposed on a large caldera that formed 27 m.y. ago (Chapin et al., 1978). Volcanic eruptions are known to have occurred along the northern margin of the caldera during three periods: 4, 7–12, and 20–27 m.y. ago. Economic deposits of silver and manganese occur within and along the northern margin of the caldera.

A number of geophysical observations suggest that dynamic processes, in addition to simple east-west crustal extension, are occurring in the Socorro area. Analysis of S-to-P and S-to-S reflections on microearthquake seismograms indicates the existence of an extensive layer of magma at midcrustal depths (18–20 km) (Sanford et al., 1977). COCORP studies in the area have provided a great deal of additional information on this magma body.

As currently mapped, the magma body has a lateral extent of about 1700 km², and the bulk of the seismic evidence suggests that it is very thin—of the order of 1 km (Chapin et al., 1978). Surface uplift roughly coincident with the spatial extent of the magma body has been documented by Reilinger and Oliver (1976), who, on the basis of an analysis of level-line data for the period 1909–1952, were able to establish average rates of uplift as great as 6 mm/yr over the magma body.

Seismic activity in the vicinity of Socorro is higher than in any other segment of the rift (Sanford et al., 1972). Although moderately strong quakes have been reported in this area since 1869, the most notable seismic activity was an intense and prolonged earthquake swarm from July 1906 to early 1907. Three shocks of this swarm were felt over areas of about 200,000 km², which suggests strong Richter magnitudes (near 6).

A characteristic of current seismic activity in the Socorro area is large numbers of small shocks, mostly occurring in swarms. The microearthquake activity is distributed diffusely over about 2300 km^2, roughly centered above the extensive magma body (Sanford *et al.*, 1977). There is no obvious relation between the earthquake hypocenters and major mapped faults in the region.

Depths of focus range from 0.5 to 13 km, with the majority between 6 and 10 km. The spatial distribution of hypocenters, along with studies of the attenuation and screening of S phases on microearthquake seismograms, suggests the existence of small bodies of magma at relatively shallow depths in the upper crust.

Heat flows exceeding 2.5 heat flow units (HFU) are commonly measured along the western margin of the Rio Grande rift (Reiter *et al.*, 1975). However, in the Socorro area, heat flows as high as 11.7 HFU have been found in the horst block mountains along the axis of the rift (Reiter and Smith, 1977; Sanford, 1977).

In summary, the important geological and geophysical characteristics of the Rio Grande rift in the vicinity of Socorro are as follows:

- Superposition of rifting on a large caldera
- Prominent intragraben horsts
- Economic deposits of silver and manganese
- An extensive layer of magma at intermediate depths in the crust (18–20 km); the possibility of small magma bodies at shallow depths in the upper crust
- High seismicity (relative to other sections of the rift), with most shocks occurring in swarms
- Heat flow as high as 11.7 HFU
- Historical uplift of the surface.

2. Geological and Geophysical Data Available

Geological and geophysical information already available on the Rio Grande rift in the vicinity of Socorro includes the following:

- Detailed surface mapping (approximately 75 percent of the area)
- Detailed gravity survey (approximately 75 percent of the area)
- Microearthquake studies (since 1960)
- Aeromagnetic survey (total area)
- COCORP data (120 km of profiling)
- Heat flow measurements (mostly shallow hole—depths less than 150 m; industry is currently making measurements of depths as great as 500–600 m)

- Magnetotelluric profile across northern part of the area
- Two refraction profiles crossing through the region
- Geochemical studies of exposed Precambrian rocks and Tertiary volcanic rocks.

3. Scientific Objectives of Drilling

The general scientific objective is to understand the dynamic processes accompanying the breakup of a craton and the formation of a continental rift. Described below are possible scientific objectives of a drilling program in the Socorro area. It is probable that drill sites could be selected that would include two or more of these scientific objectives.

a. Mechanism of crustal extension. Surface mapping and composite fault plane solutions indicate that east-west crustal extension is being accompanied by movement along numerous normal faults roughly parallel to the axis of the rift. A problem of considerable importance to the understanding of continental rifting (or Basin and Range structures in general) is whether the normal faults curve downward and become horizontal at midcrustal depths. Drill hole information on the orientation of fault surfaces to depths of approximately 10 km should resolve this question.

b. Mechanism of earthquake swarms. The mechanism of earthquake swarms is not well understood. In the Socorro vicinity, holes of moderate depth (3–5 km) could reach areas of the crust where earthquake swarms originate. Borehole measurements of rock and fluid properties similar to those proposed along the San Andreas fault (see Appendix D) would help explain the mechanism of swarming. It is also possible that drilling of this type could lead to an explanation for the association of earthquake swarms and magmatic activity within the crust.

c. Source of heat. The source of heat for the high heat flows observed in the Socorro intragraben horst has not been positively identified. Possible sources are deep circulation of meteoric water in fault zones, or magma bodies at shallow depths. It appears probable that this question could be resolved by the hydrological and geophysical information gained through drilling to moderate depths in the crust.

d. Structure of the Socorro caldera. Drilling in the Socorro area could be useful in testing models of the deep structure of calderas.

e. Interpretation of COCORP profiles. Interpretation of COCORP data for profiles crossing the Rio Grande rift is currently in progress at several institutions. Drill hole information in a few locations could be used to provide petrological data on the Precambrian rocks and to verify concepts or assumptions used in interpreting the seismic reflection data.

4. Opportunities for Add-on Experiments

A number of geothermal companies have leases in the Socorro area; two are currently collaborating on a drilling program that includes holes to a depth of 600 m. Thus the opportunity exists for add-on experiment operations in conjunction with industry.

D. Crustal Structure of the Gulf Coast and Southeastern United States (4)

The southern margin of the United States contains most of the nation's oil and gas reserves in a succession of sedimentary packages that are progressively younger from north to south.

Precambrian crust (cratonic basement) is penetrated by many wells along the northern periphery of the area, except in the deeper zones of the Anadarko-Wichita trough and the Mississippi Embayment.

The late Paleozoic overthrust belt of the Ouachita system covers the southern margin of the downwarped Precambrian craton. Thus far, wells and geophysical investigations have not delimited the extent of Precambrian crustal rocks or the areal extent of the metamorphic interior zone of the Ouachita system itself.

These zones are, in turn, covered by the southward-thickening Mesozoic-Cenozoic rocks of the present Gulf of Mexico Basin. Much of this younger history is well-known as a result of extensive exploration for oil and gas. Unknown, however, are the nature and evolution of the substrata. Are there large sedimentary basins beneath the salt beds? Is a Red Sea, northwestern Europe, Basin and Range, or other model most applicable to the origin of the Mesozoic Gulf Basin? What model is pertinent for the late Paleozoic orogenic events and their relationships to the Appalachians? How did the Precambrian crust of this region develop? Is it an accretionary or truncated margin? This list of basic questions could be easily expanded, but these illustrate our poor understanding of the evolution of the geological framework of this vast and economically critical region. Better understanding would provide a greatly improved basis for future exploration.

If meaningful geological models are to be constructed from geophysical data (such as seismic reflections, gravity, magnetic, and refraction velocities), cores and logs from a few strategically positioned deep wells must provide information such as rock age and types, densities, magnetic susceptibilities, conductivities, and thermal gradient. Without these constraints, a wide variety of hypothetical geological models will continue to be constructed in an attempt to describe the geological history of the Gulf of Mexico. But which of

FIGURE A-7 (Opposite)   Crustal structure of the Gulf Coast and Southeastern United States. A-A' is the approximate location of the crustal section in Figure A-8.

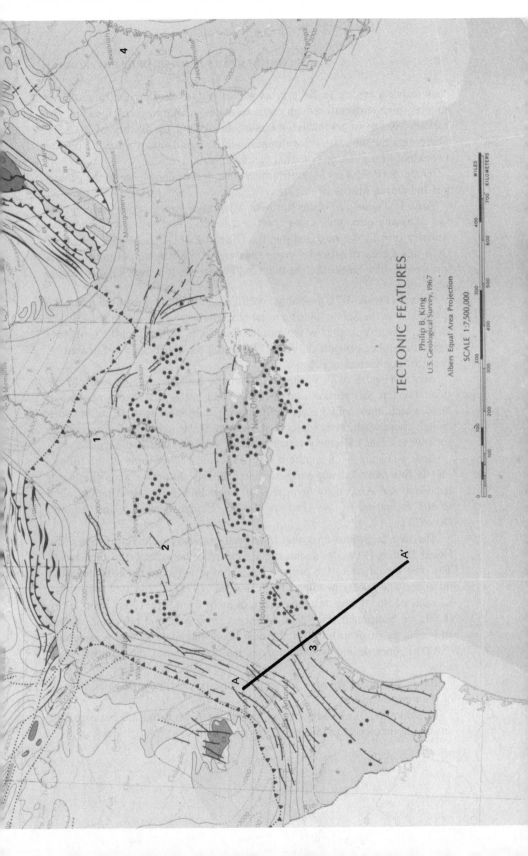

TECTONIC FEATURES

Philip B. King
U.S. Geological Survey, 1967

Albers Equal Area Projection

SCALE 1:7,500,000

these models are we to believe, as we set out to develop the potential re-sources—high-temperature/high-pressure shales, uranium, fossil fuels—in this massive volume of sediment? We must understand the history of the crust that forms the foundation for this sediment accumulation if we are to predict the extent of the energy potential contained within it.

Three deep tests could contribute a great deal of knowledge to the geological and crustal history of this area.

Area 1 (Figure A-7), the southern Mississippi trough, could help answer the following questions: Does the late Paleozoic fold belt of eastern North America turn to the west and join the Ouachitas of Oklahoma and Arkansas? Or is this a zone of offset by major shearing? If the Mississippi Embayment is a rift zone, how far south does it extend? When was it active? What basement is present?

Area 2 (Figure A-7), the Sabine uplift, could answer some of the following questions: Is the Sabine uplift a horst block between two subsiding basins, or is it a rising block that is the result of periodic vertical uplift? Is there a magma chamber beneath the uplift? When did the uplift begin in this area? Is the crust continental or oceanic?

Area 3 (Figure A-7), positioned upon the San Marcos arch to avoid the thick Louann Salt section and to understand better the history of the San Marcos arch, may help answer the following questions: Is the arch on continental, transitional, or oceanic crust? When was the arch active? What are the rock types of the extremely variable seismic velocities within the crust?

Area 4 (Figure A-7), southern Georgia, could answer some of the following: Is this line of strong magnetic anomalies the suture between the North American and African continents? If so, when did it form? How does it relate to the formation of the Gulf of Mexico, Florida Peninsula, and Atlantic Ocean?

The two deepwater tests that have been proposed in the Future Scientific Ocean Drilling (FUSOD) document are plotted schematically on Figure A-8. This test would tie to the proposed offshore tests (area 3) and lie along the refraction and gravity profile of Dorman *et al.* (1972).

Each of these three tests may require holes from 9 to 12 km deep. They should be positioned after detailed gravity, magnetic, and refraction studies and analysis of deep holes in the area. Long-record reflection seismic and COCORP lines should be obtained, at right angles, over the proposed locations.

Extremely high pressures and temperatures should be anticipated in each of the three locations.

In summary, the unknowns that could be illuminated are basic and of untold importance to the future energy picture of the United States:

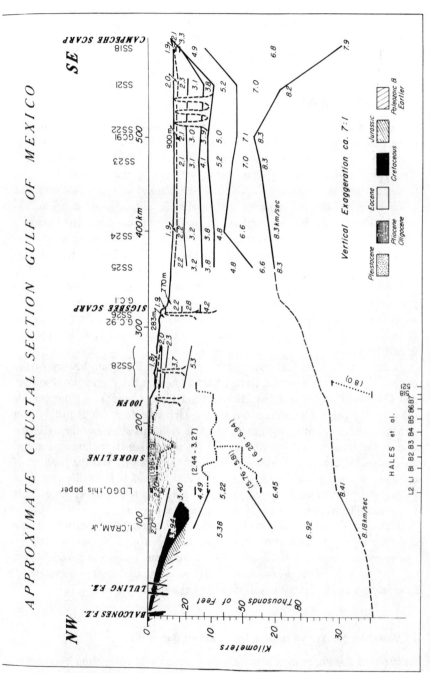

FIGURE A-8 Composite crustal section across the Gulf of Mexico approximately southeast from the Llano uplift. The data of Hales *et al.* (1970) as shown in Dorman *et al.* (1972) are projected about 160 km west to the line of section and adapted to preserve their relationship to the shoreline and the continental shelf edge. Approximate location of the section is indicated in Figure A-7.

59

- What type of crust underlies this area of interest?
- When was it formed? By what processes?
- What is the origin of the Gulf of Mexico?

E. Grenville Front (5)

The eastern flank of the North American craton was established probably by latest Precambrian or earliest Cambrian time (c. 600 m.y. ago). The last Precambrian tectonic and thermal event to affect the belt was the Grenville "orogeny," about 1000 m.y. ago; this deformation grossly altered the structure and petrology of eastern Canada and the Adirondack region, southeast of a well-defined front traceable by structural style and radiometric age southwestward across Quebec and Ontario to the Niagara Peninsula. The Grenville front is concealed southward beneath overstepping Phanerozoic strata; extrapolation across southeastern Michigan and adjoining Ohio and Kentucky is based on interpretations of regional gravity and aeromagnetic surveys. Unfortunately, neither the gravimetric nor the magnetic signatures of the Grenville event and its boundaries are particularly clear in the areas where the event is recognized by other criteria; therefore delineation of the front below Phanerozoic cover rocks is interpreted in contrasting ways by different earth scientists.

Numerous tectonic features that controlled the Paleozoic history of the eastern interior region of the United States (such as the Algonquin, Findlay, and Cincinnati arches and the Michigan and Illinois basins) lie along one or more of the postulated extensions of the Grenville front. It is not known whether, and to what extent, these tectonic elements shared in the deformation at the time of the Grenville event, or whether there were late Precambrian diastrophic controls. Further, if the Grenville orogeny was the product of continent-margin plate-convergence deformation, and if Grenville and later Paleozoic tectonics were related, then perhaps there is a plate tectonics basis for certain Phanerozoic events of the continental interior.

Such speculation is subject to testing by defining the position of the Grenville front to the southwest, below the cover rocks. No doubt some of the necessary observations are available in basement samples recovered from exploration drill holes. Resolution of the problem will require sampling of basement rocks at selected points along the extrapolated trend.

F. Basement Rocks of Cratonic Sedimentary Basins (6)

As is noted elsewhere in this report, there are several competing hypotheses to explain the subsidence of sedimentary basins of the cratonic interior. One class of such hypotheses depends on the emplacement of excess mass below basins by the filling of pre-Phanerozoic graben or rifts with relatively dense volcanics. This scenario is partially supported by the gravity field of the Michigan Basin. There a high-amplitude positive anomaly crosses the basin from northwest to southeast, and deep drilling indicates that this anomaly reflects basaltic or gabbroic rocks presumably coextensive with the positive anomaly.

Other North American basins, notably the Hudson Bay and Williston basins, share the pre-Carboniferous subsidence history of the Michigan Basin to a remarkable degree, but neither exhibits the well-defined gravity pattern of the Michigan Basin. Nevertheless, late Precambrian rifting with accompanying extrusion or intrusion of dense rocks may have been more diffusely distributed, resulting in ill-defined gravity fields typical of many basinal areas.

The Williston Basin is currently undergoing widespread exploratory and development drilling for oil and gas; many wells routinely penetrate Ordovician strata at depths approaching 10,000 ft ($\approx$ 3,000 m) and would require less than an additional 1,000 ft ($\approx$ 300 m) to reach and sample the basement rocks. The lack of geophysical resolution of basement features of the Williston Basin makes it unlikely that a single hole, or even a small number of holes, would adequately represent basement rocks, but the opportunities for relatively low cost add-on experiments, plus existing cores and samples, would make it possible to settle the question of Precambrian contribution to basin subsidence.

G. Continental Interior Basement Mapping (7)

A basic goal of continental geological research is to unravel the evolution of continents—their growth and fragmentation. This framework is a valuable resource on which to focus exploration for resources.

Much of the continental crust is covered by sedimentary rocks. In the continental interior of the United States, many exploratory holes have reached the crust. Continuing exploration for oil and gas in new regions commonly targets to the lowest expectable commercial reservoirs. These are almost always near, but not at, the underlying basement crust. Bottom-hole funds to extend these to the crust would be a valuable mechanism for obtaining samples from regions for which rock type and formation age are still unknown. As knowledge of a subsurface terrane develops, specific holes to resolve problems should be proposed to fill in the geological framework of that region. If there are no basement tests within 50 miles (80 km) of a proposed exploratory hole or if it straddles or lies on the opposite side of a recognized geophysical boundary, then it would be desirable to extend it into the basement.

H. Adirondack Region (8)

1. Introduction

Two important and related problems facing earth scientists today are the nature of the deep crust and the siting of depositories for high-level nuclear waste disposal. The Adirondack region of New York contains the deepest crustal rocks exposed anywhere within the continental United States; hence it is an excellent area in which to gain information about materials in the lower crust and processes involved in their formation. In addition, the Adirondacks contain the largest and best-studied anorthosite massif within the United States. Certain features of anorthosite, to be discussed shortly, indicate that this rock type may be a good candidate for housing nuclear waste.

2. Brief Geological Overview

The Adirondack region of New York is an extension of the Precambrian Canadian Shield into the United States. The area can be divided roughly into a lowland region in the south, west, and northwest portions and a highland region in the central and northeast portions. The highland region, locally reaching elevations in excess of 500 ft ($\approx$ 1600 m), represents an area of anomalously high surface compared with the rest of the Canadian Shield.

The Adirondack highland is dominated by a large mass of anorthositic rocks with the overall structure of a dome (Balk, 1930; Buddington, 1939). A gravity survey by Simmons (1964) indicated that this anorthosite massif is a 3- to 4-km sheet, with two roots extending to depths of approximately 10 km. The remainder of the highland consists of plutonic bodies of syenite, gabbro, and granite, and a thick sequence of supracrustal rocks of Grenville age (including marble, quartzite, amphibolite, and a variety of schists). The Adirondack lowland is composed of granite and syenite intrusions interspersed with Grenville supracrustals. Many of the igneous rocks in the Adirondacks were apparently emplaced into the supracrustals about 1100–1120 m.y. ago (Silver, 1968).

The Adirondack region was subjected to intense regional deformation and metamorphism between 1000 and 1600 m.y. ago (Silver, 1968). The present level of exposure is predominantly granulite grade, but ranges down to amphibolite grade in portions of the lowland (Engel and Engel, 1958).

Since the late Precambrian, the Adirondack region has apparently been a relative geographical high, as suggested by the onlap of Paleozoic sediments on all sides, and a geologically stable area, although it probably experienced mild regional warping during the Appalachian orogeny.

However, recent tectonism in the Adirondacks is indicated by the occurrence of microearthquake swarms centered on the Blue Mountain Lake, and by regional uplift at a rate of as much as 3 mm/yr, revealed by releveling along two geodetic survey lines. The inferred effects of Paleozoic orogenies, together with the observed recent uplift, probably account for the anomalous elevations in the Adirondacks.

3. Problems to be Investigated

The fundamental problems that can be addressed by drilling in the Adirondack region include investigation into the nature of deep crustal rocks, investigation of vertical variation of the chemistry and petrology of the anorthosite, identification of the rocks that underlie the anorthosite massif, and evaluation of the anorthosite as a potential site for nuclear waste disposal.

a. Deep crustal rocks. The nature of metamorphic mineral assemblages in many localities throughout the Adirondack highland suggests pressures of formation in the range 8–10 kbar (8,160–10,200 kg/cm^2), indicating depths of burial of as much as 30 km. Thus the rocks in the Adirondack highland represent perhaps the deepest crustal levels exposed anywhere in the United States. Consequently, a 10-km drill hole in such an area would provide an opportunity to study materials that recrystallized at depths of 30–40 km beneath the earth's surface.

Analyses of phase petrology in samples retrieved from such a drill hole should enable us to estimate the paleogeothermal conditions, whereas analyses of sample chemistry should permit us to place better limits on the composition of the lower crust. In addition, study of these samples will permit us to understand better the role of vertical chemical zoning in crustal evolution and to evaluate the effects and extent of melting processes in the lower crust. Results of these studies can be combined with data from deep crustal xenoliths found in diatremes to obtain a clearer picture of the distribution of rock types with depth in the crust.

Other studies would include *in situ* measurements of the thermal gradient and of stress. Such data may bear on the causes for the overall high elevation in this region, as well as the implications of the microearthquake swarms. For example, although near-surface heat flow in the Adirondacks is low (averaging 0.8 HFU) and heat generation is also low, do the occurrence of the microearthquake swarms and the high elevation indicate anomalous thermal and stress conditions at depth? If melting is going on at deep crustal levels, conductive pulses of thermal energy may not yet have reached the surface.

b. Studies of the anorthosite massif. The Adirondack anorthosite massif, like most other massifs, is made up almost exclusively of plagioclase (approximately 80–100 percent), with minor amounts of pyroxene, ilmenite, and garnet. Because of its nearly monomineralic character and the absence of minerals that commonly lead to the development of planar structure (e.g., biotite and hornblende), anorthositic rocks typically respond to deformation as coherent masses. Consequently, despite the great amount of tectonic activity that has affected the Adirondack region, the anorthosite itself tends to be massive, displaying little effect of penetrative deformation, and preserves igneous flow structures in many localities.

The anorthosite is highly impermeable, has low porosity, is strongly resistant to fracturing, and has a low coefficient of thermal expansion. In addition, the mineralogy is anhydrous and highly refractory, and anorthosite begins to melt at temperatures of $1100°-1150°C$ (Luth and Simmons, 1968). These physical characteristics suggest that anorthosite may be a good potential rock type in which to house nuclear waste materials.

Since waste materials would be buried at depth, more than one drill hole should be made in the anorthosite to evaluate its physical properties beneath the surface. Measurements need to be made of rock strength, permeability, fluid content, microcrack characteristics, and *in situ* thermal gradient to verify the above claims. Measurements of *in situ* stress should also be made to understand the implications of the microearthquakes for the anorthosite.

Samples retrieved from this drill hole would provide an opportunity to make detailed studies of the vertical distribution of major, minor, and trace elements in a continuous section through an anorthosite massif, and to examine vertical changes in mineral chemistry. Such information would increase substantially our understanding of the crystallization history and origin of these rocks. Finally, if the drill hole were extended to a depth of approximately 5 km, it would penetrate the rocks beneath the anorthosite. This would provide the first opportunity to study material that occurs under an anorthosite and should enable us to understand better the environment into which anorthosites are emplaced. Determining whether the underlying materials were originally plutonic igneous rock or supracrustal in origin is critical to understanding the origin of anorthosite.

In addition, drilling to a depth of at least 5 km will confirm whether the geophysical bottom of the anorthosite that is inferred from the gravity survey corresponds to a sharp lithological discontinuity. Alternatively, the anorthosite may grade (by an increase in mafic mineral content with depth) into gabbro, and such changes may be so subtle that they are beyond the resolution of the gravity technique. Only a deep drill hole can answer this question.

I. Precambrian Deep-Hole Targets in Minnesota (9)

It is becoming increasingly evident from active research in the Lake Superior region that two Archean terranes of fundamental significance are juxtaposed along a major crustal suture (Morey and Sims, 1976; Sims, 1976a). The suture is marked by a northeast-trending line of modern earthquake epicenters in western Minnesota (Walton, 1977). It passes beneath Proterozoic rocks in eastern Minnesota and coincides with the axis of the early Proterozoic Animikie Basin (Morey, 1978a). It is offset by transform faults associated with the middle Proterozoic Keweenawan rift system at the head of Lake Superior and continues across northern Wisconsin and upper Michigan as a mappable boundary between early Archean gneisses and late Archean greenstone belts (Sims, 1976a). Deep troughs in the bedrock basement beneath Pleistocene sediments in Lake Superior (Wold, 1969) may also have some relationship to this feature.

The Archean terranes north and south of the suture contrast dramatically in lithology, ages of major components, tectonic styles and history, grade of metamorphism, and mineralization. Fundamental geodynamic differences in these terranes have persisted through geological time.

The problem of the relationships between the Archean terranes is more easily visualized if the Proterozoic and Phanerozoic supracrustal terranes are swept away. Of the five Precambrian terranes recognized in Minnesota, two are basins or platform deposits of sedimentary rocks, in some respects not unlike many later Phanerozoic basins on the craton. These are the Animikie Basin of east-central Minnesota (c. 2000 m.y.) and the Sioux quartzite of southwestern Minnesota (c. 1700–1600 m.y.). Removing these along with the Phanerozoic sedimentary rocks from the geological map of Minnesota leaves the two Archean terranes split by the mafic igneous rocks within the Keweenawan rift system, which formed c. 1100 m.y. ago (Figure A-9).

Terrane I is exposed in a long strip down the Minnesota River valley in southwestern Minnesota. It is characterized by gneisses of high-amphibolite-to-granulite metamorphic facies that have had a long, complex history. Certain components of the gneiss yield radiometric ages of c. 3700 m.y. (Goldich and Hedge, 1974). There appear to be metamorphic overprints at c. 3500 m.y. and 3000 m.y. (Goldich et al., 1976). Widespread migmatization and invasions by synkinematic to late-kinematic granite occurred c. 2600 m.y. (Goldich et al., 1970). Another episode of granitic plutonism occurred c. 1800–1700 m.y., especially in east-central Minnesota (Goldich, 1968, 1972; Morey, 1977), and at about the same time, basins of volcanogenic sedimentary and volcanic rocks appear to have formed in the old gneiss terrane in northern Wisconsin and upper Michigan (Sims, 1976a). In short, the terrane has been characterized by repeated strong magmatic and tectonic activity over a period of 2 b.y.

By contrast, terrane II, north of the suture, appears to have formed during a single relatively brief episode of intense volcanic and plutonic activity c. 2700 m.y. ago. This is the terrane of greenstone and granite belts of the so-called Superior province of the Canadian Shield. It is characterized by steeply tilted and strongly to moderately deformed, but weakly metamorphosed, belts of mafic pillow lavas and volcanogenic sedimentary rocks lying between elongated granitic plutons with seemingly diapiric relationships to

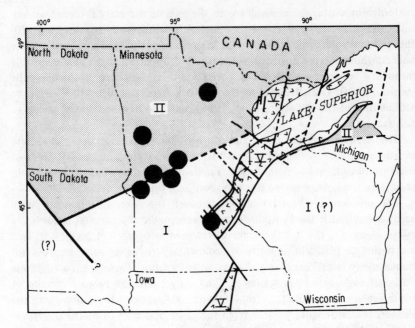

FIGURE A-9 The fundamental Precambrian basement terranes of Minnesota and adjacent areas as they would appear if overlying sedimentary strata were removed. Black circles are the locations of historic earthquakes within Minnesota. The map does not show the lower Proterozoic rocks of terrane III (Animikie Basin), terrane IV (Sioux quartzite), or rocks of Phanerozoic age.

the volcanic belts (Sims, 1976b). Numerous radiometric ages on a variety of major rock units all fall within the range of 2700 ± 50 m.y. (for a summary, see Morey *et al.*, in preparation). Metamorphism reaches the upper amphibolite facies in some gneiss and migmatite associated with granitic belts, but metamorphism is mainly within the greenschist facies (for a summary, see Morey (1978b)).

What, then, is the relationship between these profoundly different major components of the North American craton? How have they come into juxtaposition? Are they genetically related or related only by plate movements? If we knew the answers, we would know a lot more than we now do about the early history of the crust and its imprint on later geodynamic events. As it is, we can only speculate.

The boundary zone is not known to be exposed anywhere. Water well drillers' logs from a few wells in central Minnesota imply that there may be a broad zone of shearing north of the line on Figure A-9. One exploratory hole drilled last summer in connection with the U.S. Department of Energy National Uranium Resource Evaluation (NURE) program near Morris, Minnesota, at the epicenter of an earthquake that occurred July 9, 1975 (the westernmost epicenter shown on Figure A-9), encountered very weak, highly fractured granular leucogranite beneath less than a hundred meters of glacial drift (D. L. Southwick, personal communication, 1978). Both the aeromagnetic (Zietz and Kirby, 1970) and the gravity (Craddock *et al.*, 1970) maps of Minnesota show strong gradients and distinct changes in trends along the zone.

The line of suture is drawn on Figure A-9 to coincide with the epicenters of four of the seven earthquakes in Minnesota that have been reported in historic time (about 140 years). On this basis it seems reasonable to presume that at present the suture is a fault along which small movements occur from time to time, which may have interesting implications for various structural trends and linears that have been projected in younger terranes (Warner, 1978), but ideas of what and where terranes I and II were in relation to each other 2700 m.y. ago and when and how they formed the present suture remain conjectural.

It has of course been speculated that old gneisses of terrane I type formed a basement for terrane II volcanism—that remobilization and anatexis of the old basement generated diapiric granitic masses that rose upward while belts of volcanic rocks subsided in between, creating a kind of clothes-wringer vertical tectonics. Although the tectonic aspects of this model are plausible, isotopic ratios and quantitative trace element studies in the terrane II rocks of Minnesota do not favor recycling of materials that are much older than 2700 m.y. or are sialic in composition (Arth and Hanson, 1975).

Another line of reasoning speculates that terrane I represents a remnant of a very early piece of protocontinental crust whose evolution may be related to early impact processes (Weiblen and Schulz, 1978). This crust might have become rigid enough to fracture on a large scale 2700 m.y. ago, so that oceanic crust began to form with pillow lavas, island arcs, and other phenomena related to interactions between newly formed oceanic plates, as represented by greenstone belts, and older crust, as represented by gneissic terrane rocks. It is conceivable that both the oceanic and the older crustal segments were welded together by the diapiric addition of large volumes of granite during a period of major crustal thickening 2650–2700 m.y. ago (Sims and Morey, 1973). Conversely, it is also conceivable that terranes I and II were unrelated 2700 m.y. ago and drifted together later, perhaps at the time of the so-called Penokean orogeny, which generated plutonism and volcanism in those parts of the Animikie Basin underlain by terrane I rocks (Van Schmus, 1976). Terranes I and II clearly were joined together by at least 1800 m.y. ago to form a single large craton. Shortly thereafter, terrane II was elevated in relation to terrane I, because the Sioux quartzite, which occupies a basin in terrane I, has current directions and sediment composition indicating that it formed from very mature sediments derived from the north (Weber, 1977).

No doubt there are other possibilities, and the game of model building could go on indefinitely within the not too limiting constraints of present knowledge. Clearly, the ultimate answer to the nature and origin of the suture will have a strong bearing on geodynamic concepts, but equally important would be an answer to the question of what it is about the fundamental structure and composition of terrane I that has caused it to have been tectonically active over a prolonged period of geological time while the adjacent rocks of terrane II remained virtually inert except for the subsidence involved in the formation of the Animikie Basin. Two different kinds of Archean crust seem to be involved, and we need to characterize them petrologically, geochemically, structurally, and stratigraphically.

What then, from the standpoint of continental drilling programs, are the primary targets? First, in our opinion, is the need for a number of shallow holes to define the structure of the suture and the characteristics of the rocks of both terranes in contact along the suture (Morey, 1976). Glacial drift in the area ranges to several hundred meters in depth. A series of holes to bedrock with enough penetration of bedrock to get perhaps 50 m of good core are needed along at least two traverses across the suture zone. Six to ten holes are needed for each traverse, making a total of twelve to twenty shallow holes.

Second is the question of a target for deep drilling. The classical method by which geologists project what lies at depth in the crust is to develop a stratigraphic succession and project it structurally. This works in principle until mapping reaches the lowest rocks exposed in the most deeply eroded uplift, at which point there is nothing more at the surface to tell you what is below. It is arguable that this point has been reached for the North American continent in the 3000-m.y.-old gneiss in the Minnesota River valley. Further, direct knowledge of what lies at depth can be obtained only by drilling. It is also arguable that this knowledge is critical for understanding the early origins and subsequent activity of the crust.

It has been conjectured, for example, that beneath the very oldest rocks now exposed, younger rocks may be encountered as the result of crustal thickening by a process of "underplating" (see, for example, Sims, 1976a). Prolonged buoyancy is implied by the continued uplift needed for the deep erosion of the highly metamorphosed old rocks; this buoyancy in turn implies that the underplated materials are a lighter fraction of the mantle. If these younger, underplated materials exist, what might they be? Granite? Anorthosite? Whatever is found by deep drilling beneath the oldest rocks will have profound implications for geodynamic thought.

The Minnesota Geological Survey is actively engaged in a series of programs that provide a foundation for a drilling project. The data base of subsurface information is being expanded by collecting all useful water well logs throughout the state. Special attention is being given to wells that penetrate

Precambrian basement, and under the NURE program, water samples from these wells are being collected and analyzed for about 30 elements. This will provide data for geochemical mapping of the drift-buried Precambrian areas. Under a contract with the Nuclear Regulatory Commission, crustal structure and seismicity are being investigated by means of a sensitive seismic monitoring array and portable seismometers connected by telephone or microwave in real time to a central recording station. The COCORP deep seismic profiling project has given high priority to a traverse across the terrane I/terrane II boundary in Minnesota. Several hundred high-quality isotopic age determinations have been made on Precambrian rocks in Minnesota; the Minnesota Geological Survey has just completed a thorough review of these data and entered them into the U.S. Geological Survey Radiometric Age Data Bank (RADB) system. Minnesota is one of the few states with published aeromagnetic and gravity maps at a scale of 1 : 1,000,000. A program is nearly complete that will upgrade the gravity map so that most of the state will soon be covered by stations at 1-mile intervals. Recommendations are being made to the legislature as the result of a state-funded conference on geophysics to make a new high-resolution aeromagnetic map and undertake a continuing program of other geophysical investigations.

J. Continental Crust in Southern California (10)

Widespread exposures in southern California and southwestern Arizona of low-grade metasedimentary and metaigneous rocks, the Orocopia-Pelona-Rand schists of oceanic affinities, are of uncertain paleogeographical position. Most importantly, these schists everywhere underlie Precambrian crystalline basement, its cover of Paleozoic and Mesozoic sedimentary and volcanic rocks, and Mesozoic plutonic rocks. The schist, however, is rarely intruded by Mesozoic plutons. Most workers in the region regard the schist protolith to be Mesozoic in age and emplaced beneath the continental crust of southern California during latest Cretaceous or earliest Tertiary time.

The major problem presented by the geological relations is: What is the eastern extent of the schists beneath the southeastern part of California and Arizona? Some hypotheses suggest that the schist is only local, originating from a small and isolated Mesozoic paleogeographical element that was interleaved by thrusting into the Precambrian and Mesozoic crystalline crust. Other hypotheses suggest that the schist may underlie, at depths of 1–10 km, most of the Precambrian and Mesozoic crust, having been emplaced by underthrusting of oceanic rocks from the west. The resolution of these two hypotheses has tremendous societal and scientific implications. The San Andreas, San Gabriel, and San Jacinto faults cut the same windows, exposing the schist. The understanding and modeling of these faults, their earthquake activity and potential, and other tectonic effects, are strongly dependent upon the character of the crust. It is possible that the crust (30–35 km thick in this area) is composed of only 1–10 km of Precambrian and Mesozoic crystalline rocks and that the remainder is low-grade schist. The response of this type of crust to tectonic activity has not even been contemplated. Scientifically, the great extent of the schist may suggest a tectonic process not yet included in our growing catalog of continental margin tectonic processes. Geological relations indicate that the Mesozoic oceanic rocks were emplaced beneath an older crystalline crust and a Mesozoic Andean arc, by stripping away the lower half of the older crust and arc.

COCORP plans a reflection line across this terrane, and considerable mapping has been done or is in progress in the area. The geophysical work may indicate reflectors at depth, one of which could mark the contact between the older crystalline crust and the schist, but only by drilling can the existence of the schist be confirmed. Site locations for drilling should be east of the tectonic windows exposing the schist. Currently, there are three possibilities, east of the Orocopia, Chocolate, or Rand mountains. Depths to target might range from 1 to 8 km, depending upon location.

K. Third-Dimensional Relationships of Xenolith Occurrences (11)

Xenolithic rock fragments from the deep crust sampled by volcanic processes give insights as to the bulk composition, physical properties, and fluids characteristic of source regions from which some silicic intrusive and extrusive rocks are derived. In some instances, such xenoliths represent the residual material left behind following a thermal event during which the lower crust has been melted and acidic magmas have been extracted to shallower levels.

Drill holes in basement areas adjacent to xenolithic localities thus may provide information about the interrelations between source regions at depth and crystalline rocks exposed at the surface. Certain important problems may be addressed by such a drilling project, including the following: the migration of uranium-, thorium-, potassium-, and metallogenic-rich fluids into the middle and shallow levels of the crust in silicic magmas and/or hydrothermal systems; the relative abundance of such elements in different levels of the crust; and the shape and subsurface structure of the volcanic features (such as diatremes, kimberlite pipes, and maars) that serve as hosts for xenoliths, as these often are important sources of commodities such as uranium and diamonds.

A recent investigation in the Massif Central in France has used a shallow drill hole to provide additional information about the interrelations between xenoliths and the unexposed crystalline basement. This study has demonstrated the value of a drilling program that has been guided by carefully conceived objectives. Studies of crustal xenolith suites at the Bournac locality in the Massif Central were correlated with extensive mapping and geophysical surveys of the surrounding area, which contains exposures of Precambrian basement (Leyreloup et al., 1977; Daigniers and Vasseur, 1978). After compilation of these results, a site was chosen for a 200-m borehole into the basement adjacent to the pipe in order to improve the extrapolation of surface measurements at depth, understand the space and time relations of the xenoliths and exposed basement, and provide insights into the constitution, origin, and evolution of the continental crust in this area. The results of the combined surface and borehole studies demonstrated that there is a direct relation between cordierite granites exposed at the surface and the xenoliths derived from depth; in contrast to the predictions of Lambart and Heier (1967), the lower crustal granulites (xenoliths) do not show any obvious light lithophile elements (LLE) depletion; and additional information about heat flow and seismic structure provided valuable additions to the predrilling data base. Proposed drilling sites are discussed in the following sections.

1. Wet Mountains, Colorado Front Range

The core of this mountain range contains the highest-grade metamorphic rocks in the Cordillera; many of these rock types are similar to xenoliths scavenged by kimberlite diatremes farther north on the Colorado Plateau (McGetchin and Silver, 1970). In addition, the core complex contains alkaline intrusives with certain known deposits of thorium, rare earth elements (REE), and vermiculite, and a layered mafic-ultramafic intrusion with associated iron and titanium deposits (Shaw and Parker, 1967; Singewald, 1962; Boyer, 1962). The metamorphic core complex provides a particularly fruitful drilling site for the following reasons: (1) A hole 5,000–10,000 m deep in the metamorphic core complex (Figure A-10) will provide a basis for correlating the geological and geophysical information (thus derived) with what is known about the crust from xenolith studies, geophysical surveys, and mapping of the exposed crystalline basement. This will result in a better understanding of the source regions for the uranium-bearing silicic magmas, which are the origin of economic deposits of uranium on the Colorado Plateau. (2) Such drilling may also indicate the presence of additional alkalic or mafic-ultramafic layered intrusions of potential economic importance. (3) The drill hole has the potential of improving our understanding of faults responsible for the vertical tectonics that control exposures in this region, particularly if the hole penetrates the base of a thrust plate. (4) It may be possible to calibrate and extend the crustal model of McGetchin and Silver (1970) into the region.

2. Southern Rio Grande Rift Adjacent to Kilbourne Hole Maar and Potrillo Maar in the East Potrillo Mountains Horst

A major crustal discontinuity (the Texas lineament?) occurs across the southern Rio Grande rift. Totally different crustal xenolith populations occur at

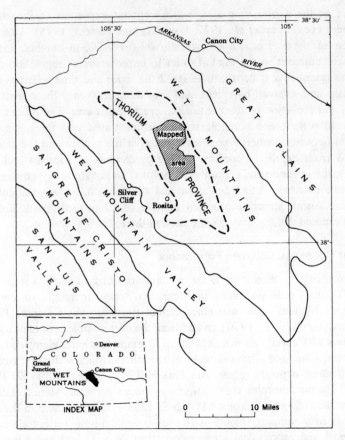

FIGURE A-10   Location of thorium deposits, Wet Mountains,
Colorado.

Kilbourne hole and Potrillo maar, which are only 18 km apart (Padovani and
Carter, 1977). Kilbourne hole maar reveals an anhydrous lower crust, whereas
primarily hydrous greenschist facies metamorphic rocks are present at Potrillo
maar. No greenschist facies metamorphic rocks have been found at Kilbourne
hole. This suggests a fundamental difference in the crustal blocks between
these two very close maars. A deep hole of the order of 10 km near each maar
(Figure A-11) should allow calibration of the xenolith model for the lower
crust and resolution of the basic question of the evolution of the crust in this
region. A single hole near the Potrillo maar is potentially the most productive.
The hydrous crustal rocks at Potrillo may represent fragments of a thrust
sheet from Mexico that terminates against the Texas lineament. In addition,
the observation that the oldest ash deposits at Potrillo contain mantle-type
xenoliths, whereas the younger deposits contain crustal xenoliths, implies a
multistage eruptive process in which the magma "sat" at different levels

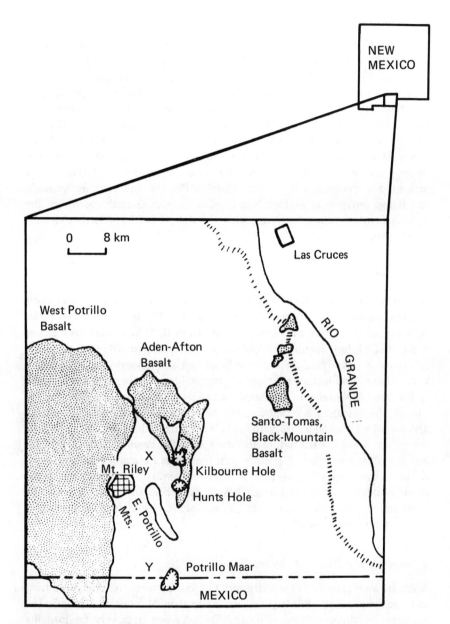

FIGURE A-11   Location of the Kilbourne hole–Potrillo maar area of south-central New Mexico (Hoffer, 1975).

within the crust (Carter, 1977). Additional information pertinent to the evolution of magma chambers and to geothermal resources would be derived from this study in a known geothermal resource area (KGRA).

L. Columbia Plateau (12)

Several important and interesting problems involving deep continental structures in the Columbia Plateau may be solved only by deep drilling. In a broad sense, the Columbia Plateau province may be divided into three completely different areas—the Snake River Plain, the Columbia Plateau proper (Oregon and Washington), and the Blue Mountains. Also included in this discussion is the northern end of the Basin and Range province in southeastern Oregon. The problems and importance of the deep basement structure beneath the Snake River Plain are presented elsewhere in this report. However, two other major problems are concerned with the nature of the shallow (10 km) crust beneath the southern half of the Columbia Plateau, and beneath the Basin and Range province in southeastern Oregon. In both areas the nature of the rocks beneath the lowest exposed unit, the Columbia River (Owyhee) basalt, is unknown.

1. Columbia Plateau of Oregon and Washington

Several relatively deep holes have been drilled in the Columbia Plateau basalts (approximately 1.5–1.8 km deep), and one, the Rattlesnake Hills well, has been drilled to a depth of 3 km. The data from these holes have been extensively studied, but significant problems are still unsolved. The nature and age of the rocks underlying the Miocene basalt are still unknown. A 1.8-km hole near Odessa, Washington, bottoms in granite. Mesozoic basement is exposed in the Blue Mountains of northeastern Oregon. Between these areas the nature of the basement is unknown, as is the thickness of the Miocene basalt. The questions of interest are as follows: What are the thickness and the age of the basalt? What are the nature and the age of the basement? Was the basalt laid down on a remnant of oceanic crust present as a basin or a fault block or on an existing Mesozoic continental basement similar to that exposed to the north and south or along a major NW-SE fault zone of crustal dimensions? What are the temperatures, hydrology, and physical and chemical properties of the upper crustal rocks?

2. Southeastern Oregon Basin and Range

A similar problem exists in southeastern Oregon. Between the Blue Mountains and near the Nevada border, no rock older than Miocene is exposed. The basement is Miocene Owyhee basalt. No holes penetrate very far into the basalt, although one or two deep wells have been drilled in the basin areas. The problems to be addressed are similar to those noted above: What are the thickness and the age of the basalt? What are the nature and the age of the basement below the basalt? Is it Mesozoic rock, or a Cenozoic basement formed by back-arc spreading, or a combination of the two? What are the temperatures, hydrology, and physical and chemical properties of the upper crustal rocks?

In general, the objectives of drilling in both areas will be to furnish key data relating to the origin of the major changes in structural trends of the Mesozoic rocks in the Klamath–Blue Mountain trends; information on the plate tectonics evolution of the Pacific Northwest; data on deep basement structure; and an assessment of the deep geothermal, hydrological, and mineral resources that may exist beneath the basalt cover.

M. Thrust Fault Geometry and Its Relation to Basement (13)

Decollement thrust faults, which separate sedimentary cover rocks from their crystalline basement, are an integral part of most mountain belts that have been recognized for more than 100 years. As thrusts of this type are traced toward the central part of a mountain belt, their geometry becomes complex and ill defined, because either the rocks become progressively more highly metamorphosed or else older basement rocks appear in structurally high positions and their relation to the decollement thrusts is uncertain. The geometry of the decollement thrust faults, in their more internal parts, is critical to our understanding of their mechanics, and better understanding is required to place better constraints on existing models of thrust faulting. The main questions to be addressed concern the three-dimensional geometry of the internal parts of decollement thrust faults, and the relation of these thrust faults to contemporaneous metamorphic core areas, or the structurally older rocks in high positions. Better knowledge of the geometry can then be related to the mechanical questions of the role of gravity and crustal-shortening processes in the formation of decollement thrust faults, and their possible relation to plate margin mechanics.

Thrust belts have been extensively mapped, and oil companies have provided excellent seismic profiles to demonstrate their three-dimensional geometry in the external foreland or in the continental side of the thrust belt. Traditionally, oil companies have stopped their seismic profiles where metamorphism begins in the rocks or where structurally high basement appears. Geological relations suggest that many of the metamorphic areas or structurally high basement rocks are themselves thrust, and foreland decollement thrusts pass beneath them. If this is true, exploration targets may underlie the more internal (hinterland) regions of thrust belts, heretofore unexplored. Our current societal demand for energy requires that we take new and bold approaches to the exploration for oil and gas, and the internal part of thrust belts is a logical prospect.

The basic problem of the geometry and mechanics of thrust faulting is of international scope. Most mountain belts in the world contain such structures. Our present three-dimensional model for these structures, and their relation to metamorphic cores and structurally high basement, comes from piecing together surface exposures from different structural levels exposed at different places in mountain belts eroded to different levels. This view is not entirely satisfactory, because we can never really be sure that each piece of information used in the geometric construction comes from comparable examples or that we are dealing with separate parts of the same "elephant." Good geological, geophysical, and drill hole data on a single example may provide a complete geometry or a case history that might serve as a world standard.

Proposed drilling sites may be envisioned in the Appalachians and the Cordilleran Mountain belts. The Appalachians contain several excellent sites, where structurally high basement is present in the internal part of the thrust belt. The southern part of the Blue Ridge province along a COCORP line may be potentially the best. Other sites could be the Blue Ridge anticlinorium in Maryland or the Berkshire Highlands in eastern New York State. The southern Blue Ridge and Berkshire Highlands area would probably require the shallowest holes, perhaps approximately 3–5 km deep, whereas those at the Maryland site might need to be considerably deeper.

The Cordilleran Mountain system contains several potential sites that have advantages over the Appalachian ones in that they may be related to a reasonable plate tectonics interpretation. Southern Idaho offers the best possibility, as the thrust belt there is well mapped, has good industry seismic profiles in its eastern part, and has been a model for thrust tectonics in the Cordillera; western Montana–eastern Idaho, near the Idaho batholith, may be another. Both examples may require 3- to 5- or perhaps even 7-km holes. A possible site in southeastern California is the Halloran Hills, where shallower holes of 500–2000 m would give slightly less complete, but very important, results.

N. Stable-Interior Crystalline Terranes—Stress, Rheological, and Petrological States (14)

Major crystalline terranes of the continental interior are of considerable current interest for at least three reasons:

- Large granite massifs are expressed as arches and domes (e.g., Wisconsin arch) that have a Phanerozoic history of slow subsidence or episodic uplift relative to adjacent sedimentary basins. As knowledge of basins and their basement rocks and structures increases, it becomes important to have equivalent data on positive elements, so that comparisons and contrasts may be drawn and interpretive models of the dynamics of arch/basin couples developed.

- Deep granitic terranes are under increasingly serious consideration as potential sites for radioactive waste repositories. Further consideration requires hard data on the ambient stress field, temperature and heat flow, detailed mineralogy, and rock mechanics properties of ancient granites below the depth of open fractures, and of accompanying groundwater penetration. Ideally, ultradeep boreholes enlarged at depth by remote mining technology, could provide self-sealing repositories that would meet stringent tests of acceptability.
- The temperature and radiometric heat production in the crust are of considerable importance in the modeling of geodynamic processes. How much of the surface heat flow is produced in the crust, and how much comes from the mantle? Present models assume an exponential decrease of heat generation with depth, but these are only approximations. Xenoliths, brought to the surface by volcanic eruptions, sometimes yield information on heat generation, but these are likely to reflect anomalously large heat sources. Only deep boreholes in regions of stable basement can provide definite information on the average heat generation in the earth's crust.

Properly controlled, adequately instrumented, logged, cored, and sampled drill holes designed for repository site investigation should satisfy many of the scientific inquiries of those concerned with the history and dynamics of stable-interior crystalline terranes. Drilling depths would probably exceed 5 km, but should result in minimal technical difficulties. Site selection, it is assumed, would follow intensive COCORP-type seismic profiling and other geophysical and geological studies.

O. Precambrian Crustal Discontinuity in Southeastern Wyoming (15)

The problem of Precambrian continental growth and the efficacy of plate tectonics in the Precambrian are represented by a major Precambrian discontinuity that is exposed in southeast Wyoming. This discontinuity is a shear zone several kilometers wide that separates two major Precambrian provinces of different age and petrography. North of the shear zone is the Wyoming province, dated 2.7 b.y. old or older. This represents some of the oldest core of the North American continent. These rocks consist of metasedimentary rocks, migmatites, and granites. South of the shear zone are rocks approximately 1.8 b.y. old and younger. These consist of gabbros, amphibolites, biotite-plagioclase gneisses, and granodiorites. The shear zone, which is several kilometers wide, consists of mylonites and cataclastic migmatites. Maximum strike-slip displacement is uncertain but measures in kilometers. The shear zone is exposed for about 150 km in southeast Wyoming uplifts and may be followed in limited exposures for another 100 km to the northeast. Evidence suggests that this shear zone continues in the subsurface across most of the western half of the United States; therefore it represents the major Precambrian crustal boundary in the United States. The shear zone itself, which may have been partially melted by frictional heating to form migmatites, has undergone chemical alteration from migrating fluids. The shear zone could be a deeper analog of the San Andreas fault.

The younger crust south of the shear zone may have been accreted to the older crust as island arcs. The composition of these rocks would be appropriate for a deeply eroded island arc. Alternatively, because of the large size of the southern area, it may have been a continental block that collided with the Wyoming province. In either instance the shear zone was a discontinuity that was later marked by transcurrent movement.

1. Present Status

Although the general features of the shear zone and surrounding rocks are known, sufficient detail for a real understanding of the area is lacking. Detailed geological and regional geophysical studies are being started, primarily to determine whether the structure may be related to a plate tectonics framework.

Drilling should be conducted with two separate goals. In the Great Plains, exploratory drill holes should be extended into the Precambrian to determine the eastward extension of the shear zone. Eventually, some deep holes should be drilled in Precambrian blocks on either side of the shear zone and into the shear zone itself.

2. Results from Drilling

Extending holes into basement of South Dakota would trace this crustal boundary to the east. Deep drilling would reveal the nature of the shear zone and the mechanism of movement at depth. This could help solve a major problem regarding earthquake mechanisms on active transcurrent faults, like the San Andreas. The role of fluids and partial melting, which appear to have been important along this shear zone, could be a critical factor in the movement mechanism. Deep drilling on either side of the shear zone would show the composition of the deep crust and its vertical variations. More important would be the detection of large crustal slices or wedges that might be expected along a collision zone—possibly also a migmatitic zone mobilized by the collision. On the other hand, drilling on the south side could reveal parallel thrusts of an island arc.

The area offers so many scientific and practical rewards that it constitutes a prime problem for deep drilling after it has been studied in more detail. The scientific rewards would include a better understanding of deep fault zones and of continental structure and growth. The southern area is a metallogenic province for base metals, associated with mafic intrusions and possibly deep zones of island arcs. The northern area is currently undergoing very active exploration and drilling as a uranium province in metasedimentary rocks. The Department of Energy is supporting some research here. The entire area provides a most unusual combination of adequate exposures for geology and adequate accessibility for geophysics to permit the solution of one of the major Precambrian problems in the United States.

P. Deep Crust of the Western Great Basin (16)

A major question that merits study through deep drilling is the structural origin and the extensive mineralization of the western Great Basin.

The results of the marked extension caused by normal faulting, as well as the complex and widespread volcanism that preceded and accompanied that extension, are not adequately understood, despite recent debate in terms of plate tectonics and plume-crust interactions. From the standpoint of continental evolution, the processes by which a possible preexisting cratonic crust may have been disrupted, and the manner in which previous crustal structure and constitution may have controlled this disruption, are of broad scientific interest. In addition, the cause of the newly discovered Battle Mountain regional heat flow high, the cause of the unique lead-isotope signatures of ore deposits in this region, and the question of whether Precambrian cratonic basement exists west of the Ruby Range are other problems in the region that can be addressed only by drilling.

From the standpoint of mineral resources, the great distance separating probable Laramide and younger subduction zones from some of the major mineralized regions of the West is one of the enigmas in current genetic theory of Cordilleran hydrothermal ore deposits. Moreover, because known mineral deposits are located in the ranges, more knowledge of the structure, constitution, alteration, and mineralization of the basin blocks is definitely needed. For example, are the graben blocks affected by greater or lesser intrusive activity and by the accompanying mobilization of metals?

Major questions exist about (1) the nature of epithermal mineral deposits in the western Great Basin, (2) heat flow in the Battle Mountain High, (3) playa sediments as economic reservoirs for lithium, boron, potassium, mercury, and other heavy metals, (4) the nature of blocks beneath the basin fill, and (5) the existence of Precambrian basement. Therefore there are abundant opportunities to drill multipurpose holes; of course, no one hole can be expected to answer all these questions. Initially, a program of two holes is suggested, one relatively shallow 3-km hole in a structural high and another approximately 4 km deep with intermittent coring within a graben to basement in the trapezoid-shaped area with corners at Battle Mountain, Austin, Rochester, and Winnemucca in west central Nevada.

Q. Wichita Mountains, Oklahoma (17)

A number of petrological and tectonic-structural concepts of this aulacogen could be tested with a small number of holes to depths of 10,000 ft (3,000 m), one of which should be cored extensively.

Drilling sites suggested are in 36-mi^2 townships based on the Indian Meridian ($\approx 97°5'W$) and the Oklahoma Baseline ($\approx 34°30'N$):

- T6N-R21W: A hole beginning in the surface outcrop of Reformatory granite would demonstrate its nature and structure and the depth to gabbro and would provide samples for detailed chemical study of both the granite and the gabbro.
- T3N-R15W: A hole with similar objectives in Quanah granite would supply analogous information.
- T1N(or T1S)-R16W: A hole beginning in the Permian would penetrate the Navajoe Mountain basalt group and perhaps the Tillman metasedimentary group and would possibly reach the underlying gabbro, clarifying the relations of these three groups, which are currently problematical.
- T4N-R17W: A hole to penetrate the Cold Springs "granite" and sample the underlying gabbro would be similarly valuable.

Many detailed arguments should be added, such as: The phase chemistry of a long section of the Glen Mountains layered complex is necessary before the size of the body can be estimated with confidence. The two coarse-grained (Reformatory and Quanah) granites appear to be sills, but estimates of their thickness and variation in uranium concentration are entirely speculative at present. Certainly, two of the named subsurface groups, the Navajoe Mountain basalt group and the Tillman metasedimentary group, will require considerable petrographic, chemical, and radiometric study before their place in the evolution of this part of the continental crust is clarified. Drilling the foregoing sites may provide the basis for answering some of these questions.

R. Geodynamics of Iceland

1. Significance for Geodynamics

Iceland occupies a unique position in the plate tectonics theory because of its position at the boundary of the North American plate on the axis of the midocean ridge, underlain by shallow asthenosphere (Palmason and Sae-mundsson, 1974). The ridge is a spreading plate boundary, where material from deep within the earth wells up to form new oceanic crust. Iceland, the only large landmass on the ridge, is easily accessible and therefore attractive to most investigations, including those best attacked by deep drilling. Scientific investigations there are important because Iceland provides a highly relevant model for the architecture, composition, physical state, and behavior of the ocean floor, fragments of former ocean floors accreted within continental lithosphere, and transition zones between oceans and continents. Major scientific objectives include the following:

a. The entire plate tectonics concept is dependent upon the formation of new lithosphere; the relationships among changes in physical and chemical attributes of the lithosphere and rates of horizontal and vertical movements seem best defined here. Yet shallow stress measurements made by stress relief methods at several locations in Iceland appear to have demonstrated significant large horizontal compressive stresses (Hast, 1973); these stresses are difficult to interpret from the standpoint of conventional plate tectonics models. A basic problem to be solved therefore is the three-dimensional stress field in Iceland, which leads directly to the question of how this stress field is related to the evolution of the midoceanic ridge and, in general, the floors of the oceans. A detailed stress measurement profile in one or more suitably located deep holes might shed important light upon this problem.

Stress measurements outside the Iceland rift zones suggest a relatively strong component of crustal compression ($\sigma_{Hmax} = \sigma_2$ or σ_1) roughly normal to the axial rift zones. Considered alone, such data might be taken as an indication that the lithospheric plates are not being forced apart. However, at least in southwest Iceland, focal mechanism solutions suggest the contrary view—i.e., crustal extension is occurring across the axial rift zone ($\sigma_{Hmin} = \sigma_3$). Both sets of data seem to be correct indications of the crustal stress environment at different physical locations; the problem is to reconcile them.

b. Plate spreading is effected in large part by normal and strike-slip faulting in the axial zone. The actual geometries and movement paths associated with these faults are known directly only from surface and near-surface exposure. But complications are indicated by focal mechanism solutions. Furthermore, the growth-fault nature of the normal faults suggests that displacements must increase downward, beyond the reach of direct surface observation. Perhaps the faults flatten with depth, but as yet no data are available. Drill-core information on the flanks of the axial zone would be essential for the solution of this important question.

c. The Deep Sea Drilling Project (DSDP) and related marine geology investigations have profoundly increased our understanding of the properties of the seafloor. But certain kinds of information available from land drilling in Iceland would do much to assist our understanding of the basic data obtained in oceanographic investigations.

Outstanding among the questions related to geodynamics of Iceland is the identification of crustal layer 3 under Iceland, which has been identified throughout the ocean basin by seismic studies. The layer is not exposed and is below the depth of DSDP drilling; therefore its true nature is not known, although the question has been subjected to considerable speculation. Layer 3 extends under Iceland, where it has been identified by numerous seismic traverses ($V_p \approx 6.5$ cm/s). Its upper boundary has been mapped in detail throughout Iceland (Palmason, 1971) and occurs mainly in the depth range of 2–5 km. (All sites where layer 3 has been recorded at depths less than 2 km are associated with central volcanoes). It is likely that layer 3 consists of complex, low-porosity, low-permeability intrusions (Fridleifsson, 1977). But identification of the physical and chemical attributes of layer 3 based on deep drill samples would end many controversial aspects of the problem and would place many oceanic and transitional continental margin geophysical studies on more secure foundations.

2. State of Existing Geological and Geophysical Knowledge

The relation of Iceland to the Mid-Atlantic Ridge has been recently and ably reviewed (Palmason and Saemundsson, 1974; Kristjansson, 1974). The subject is of much interest and has been relatively well explored, for if ocean floor spreading is taking place in the manner portrayed by plate tectonics, then Iceland must be splitting apart. The active zone of the Reykjanes Ridge segment (strike N40°E) of the Mid-Atlantic Ridge comes ashore at the tip of the Reykjanes Peninsula; the zone crosses Iceland in a relatively complicated manner, with two branches in southern Iceland, and finally connects along an oblique offshore zone with the Kolbeinsey (Iceland–Jan Mayen) Ridge.

The axial zones of rifting and volcanism contain a great variety of volcanic and intrusive forms, cut by numerous faults and open fissures running mainly NE-SW in southern Iceland, with a more northerly trend in northern Iceland. On the Reykjanes Peninsula the general trend of the axial rift zone is about N70°E (Figure A-9; Kjartansson, 1960). Individual eruptive and open fissures and faults, however, strike about N35–45°E and are thus en echelon to the general trend (Jonsson, 1967; Kjartansson, 1960; Nakamura, 1970). The nearest normal faults to borehole sites used by Haimson and Voight (1977) in a

FIGURE A-12 (Opposite)   Icelandic stress measurement sites and localities mentioned in text. Geologic base map after Saemundsson (1974). Reykjavik hydrofracturing site, R. Overcoring sites: Sandgerdi, on Reykjanes Peninsula, S; Burfell, B; Akureyri, A; Hvalnes, H; Stokksnes, S. Borgarfjördur earthquake location, Bo. Hypothetical plume centers at Tungnafellsjökull, T, and Kverkfjoll, K. (After Haimson and Voight, 1977.)

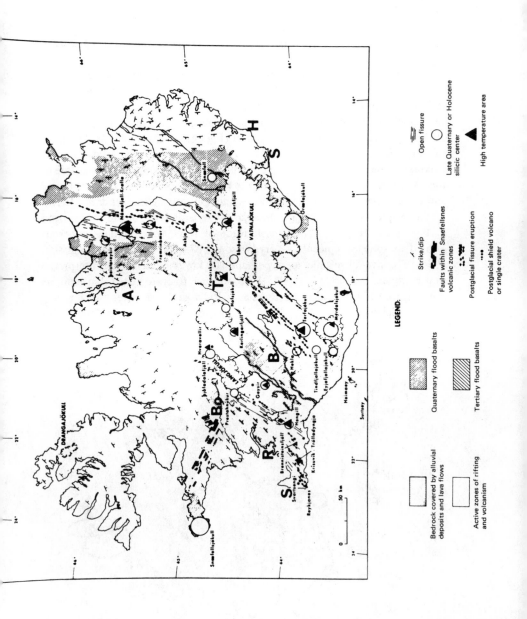

LEGEND:

Bedrock covered by alluvial
deposits and lava flows

Active zones of rifting
and volcanism

Quaternary flood basalts

Tertiary flood basalts

Strike/dip

Faults within Snaefellsnes
volcanic zones

Postglacial fissure eruption

Postglacial shield volcano
or single crater

Open fissure

Late Quaternary or Holocene
silicic center

High temperature area

0   50 km

pioneering series of preliminary borehole stress experiments (about 6 km to the southeast) strike N45°E. The general trend of the axial rift zone then undergoes a profound change to about N35°E, parallel to the trend of individual fissures and faults. This trend is maintained from south of Hengill, 30 km east of Reykjavik, to Langjokull (Figures A-12 and A-13). The measured rate of subsidence in the axial rift zone is estimated at about 4-6 mm/yr maximum, decreasing toward the boundaries of a 30-km zone on each side of the axis. Precise distance measurements suggest a combination of left-lateral and extensional movement on the Reykjanes Peninsula (Brander et al., 1976).

The Reykjanes Peninsula and an adjacent zone due east of the abrupt bend in the axial rift zone, below the latitude of Reykjavik, contain numerous located earthquake epicenters (Tryggvason, 1973; Björnsson and Einarsson, 1974). The overall seismic zone, perhaps 100 km wide, has been interpreted as a complex fracture zone (viz. the Reykjanes fracture zone) that exhibits transform motion between the Reykjanes Ridge and the eastern axial rifting zone of Iceland (Figure A-13; Ward et al., 1969; Ward, 1971, p. 2999). Mo-

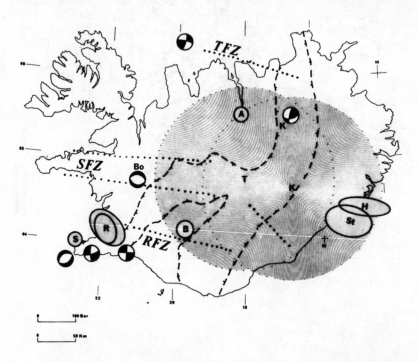

FIGURE A-13   Tectonic map of Iceland indicating observed stress conditions. Axial rift zone boundaries denoted by dashed heavy lines with barb. Tjornes (TFZ), Snaefellsnes (SFZ), and Reykjanes (RFZ). Fracture zone boundaries indicated by dotted lines. Another possible fracture zone boundary passes through Oraetajokull near the southeast coast.

tion is not simple, and on the Reykjanes Peninsula the plate boundary apparently has both transform fault and spreading ridge characteristics (Klein *et al.*, 1973, 1977; Björnsson and Einarsson, 1974; Brander *et al.*, 1976). An equivalent seismic zone off the north coast (Tjornes fracture zone) has been similarly interpreted in reference to transform motion between the eastern axial rifting zone and the Kolbeinsey Ridge (Sykes, 1967; Ward, 1971, p. 3001); the concept has found substantial support in field investigations (Saemundsson, 1974). A third fracture zone has been suggested at 64°40'N for the Snaefellsnes-Vatnajokull trend (Sigurdsson, 1970; Schäfer, 1972).

The axial rift and volcanic zones are flanked symmetrically by primarily Quaternary volcanics, which in turn are bordered by tertiary volcanics (16 m.y. maximum age). The strata in general dip toward the active zone, suggesting continuous volcanism and crustal spreading during the last 7 m.y. (Fridleifsson, 1973; Saemundsson and Noll, 1975; McDougall *et al.*, 1977).

Thermal gradients to the west of the volcanic zone in southwest Iceland increase regularly from about 70°C/km in Tertiary rocks 100 km west to about 165°C/km in early Quaternary rocks 20 km west (Palmason, 1973, 1974; Tomasson *et al.*, 1976). Variable thermal gradients, including trend reversals, are observed in the Quaternary strata because of increased water circulation.

3. Borehole Studies Preliminary to Deep Drilling

Stress measurements by the overcoring technique in shallow boreholes were performed by Hast in 1967–1968. The deepest measurements (about 50 m) were made at Burfell, 10 km west of the volcano Hekla. The most important aspect of these measurements is the indicated prevalence of compressive stress. Few directions (perhaps none) provide reliable indication of crustal stress directions at depth (Haimson and Voight, 1977, p. 174). The two possible exceptions involve two shallow (30 m) sites in eastern Iceland (Figure A-13).

In 1976, stress measurements by hydrofracturing were carried out to about 0.4-km depth in two boreholes in Quaternary volcanic rocks in Reykjavik on the flank of the Reykjanes-Langjokull continuation of the Mid-Atlantic Ridge (Haimson and Voight, 1977). Further hydrofracturing studies by the same investigators during the summer of 1978 involved preexisting boreholes at Akranes and Burfell, in southwest and south central Iceland, respectively. This series of investigations, involving add-on experiments in existing holes, has performed an important function in that various borehole measurement procedures could be examined and refined in relation to the special and often severe conditions imposed by the Iceland crustal environment. The experience thus gained is helpful in the efficient design of future deep-hole measurement attempts.

The hydrofracturing stress measurements at Reykjavik suggest a dominant regional orientation of σ_{Hmax} approximately perpendicular to the axial rift zone (Figure A-13). This orientation is supported by the more reliable of the shallow overcoring measurements in southeast Iceland and by recent earthquake focal mechanism solutions for the Borgarfjördur intraplate area. The hydrofracturing measurements indicate a depth dependence on stress environment, with associated changes in principal stress orientation. The details of stress reorientation will, however, require deeper measurements. The flanks of the axial rift zone appear to be characterized mainly by trajectories of σ_{Hmin} parallel to isochrons. The associated principal stresses may vary as a function of spatial position relative to latitude, longitude, and depth, and large lateral compression may exist, particularly at shallow crustal levels.

This state of stress is fundamentally different from that in the axial rift zones themselves. In the rift zones, σ_{Hmin} is consistently aligned perpendicular to individual rift zone fissures and faults; σ_{Hmin} may thus be oblique to the gross orientation of the rift zone itself, where (as on the Reykjanes Peninsula) the zone seems best described as an obliquely spreading ridge or a leaky transform fault.

The existing data may be consistent with the hypothesis that thermal stress components are introduced as a function of spreading, due to both axial zone lithospheric accretion and cooling, and basal lithospheric accretion and cooling. Some attempts have been made to model this effect with finite element techniques.

Further studies in Iceland are clearly necessary to test and expand upon these interpretations. Existing data do not permit a complete assessment of possible smaller-scale influences on the stress field. Further hydrofracturing studies to provide answers to some of these specific questions are needed. The Haimson and Voight (1977) study was restricted to 0.4-km depth simply because of available drilling equipment; future studies should make more complete use of the much greater depth available in existing boreholes and should employ new boreholes drilled specifically for geophysical tests. Such future studies could provide invaluable information on lithospheric stress variation as a function of surface position and depth. The implications of the Icelandic data are, as many scientists have demonstrated, far reaching.

Considerable experience in deep-drilling procedures in Iceland has been accumulating, principally through the geothermal exploration and production efforts of the National Energy Authority. Numerous drill holes at 2 km, and several in excess of 2 km, have been drilled (Palmason et al., 1975). Bottom-hole temperatures vary in depth, with maximum values of about 340°C. Normally, cuttings are recovered, with local core intervals taken as required to test specific hypotheses. Until 1975 the depth capacity of the National Energy Authority drill rig was about 2 km. This has recently been extended to 3.6 km. The deepest drill hole to date is 2.8 km, in Tertiary rocks of northern Iceland. During the summer of 1978, two holes were drilled in Reykjavik with a target depth of 3–3.5 km, about 2–2.5 km below the Matuyama-Gauss polarity transition.

Finally, the current efforts of the narrow-bore internationally funded deep-drilling project in Reydarfjördur should be mentioned. A design depth of 1.3 km is anticipated for a 75-mm-diameter hole, with an added kilometer depth planned for reduced bore (60 mm). Continuous core is being obtained, with valuable recording of magnetic polarity. The narrow bore restricts certain types of borehole studies and the maximum drill depth obtainable, but the experience gained will be invaluable in deep-penetration efforts.

4. A Candidate Site for Deep Drilling

The following strategy conditions have already been met:

- Regional geophysical and geological studies have been carried out in sufficient detail to permit recognition of suitable candidate drill sites.
- A fairly comprehensive program of shallow-hole investigations of various kinds has been completed, and additional studies are now in progress.

Candidate sites for deep drilling, for several purposes, have been considered over a period of several years by various individuals and indeed by a specific committee established for that purpose in Iceland. The sites may be classified by the following criteria: lava piles versus central volcanoes; active zone of rifting and volcanism versus "inactive" zone; and finally by temperature gradient.

For the purpose discussed herein—i.e., regionally significant stress measurements, crustal structure, and lithostratigraphy and magnetostratigraphy— the most suitable candidate sites are those that involve the lava pile in relatively cool rock at some distance from the active zone.

The Borgarfjördur area of western Iceland can be specified as an example of a suitable location south of the observed faults associated with the Snaefellsnes fracture zone. The oldest rocks in the area outcrop in an anticlinal flexure, the axis of which can be traced from Borgarnes to Hredavatn. A candidate drill site is suggested near the axis of this flexure in rocks subjected to mesolite-scolecite zone metamorphism, with the specific location established by comprehensive but short-term geophysical, geological, and terrane analysis. The drill would thus penetrate rocks that are older than those exposed on the surface in this region, penetrating at least to and preferably well within layer 3. The age of the surface rocks in the area is known from what is perhaps the most intensive paleomagnetic and K-Ar age analysis thus far carried out (McDougall et al., 1977), involving 400 successive lavas indicating a record of virtually continuous volcanism between about 7 and 2 m.y. ago.

The desired borehole depth would be in excess of 4 km, with extraction of continuous core. Samples would be subjected to (1) standard petrographic and geochemical (including isotopic) study, with analysis of the relation of lithology to secondary mineralization; (2) paleomagnetic analysis; (3) fracture and fault analysis, including study of mineralized fillings and coatings; (4) material property evaluation (elastic, seismic, strength, electrical, magnetic, and fluid penetration). The nature of bedding and fault structure would be traced through the upper boundary of layer 3. Principal measurements would include *in situ* stress via the hydrofracturing method at perhaps 10 intervals, temperature and heat flow, and geophysical logging (standard instruments, e.g., gamma-gamma density, neutron, sonic induction, resistivity, dipmeter, nuclear magnetism, and fracture orientation). Instrument modification may be required for the lower portion of the hole, in view of anticipated bottom-hole temperature.

Estimated drilling cost of the program is $2 million (1978 dollars), with an additional $1 million associated with costs for coring and scientific investigations.

Appendix B
Thermal Regimes of the Crust, Particularly Those Related to Hydrothermal and Magmatic Systems

B THERMAL REGIMES OF THE CRUST, PARTICULARLY THOSE RELATED TO HYDROTHERMAL AND MAGMATIC SYSTEMS

Heat stirs the restless earth. Ultimately, it drives the plates and determines the sites of concentration of ores and geothermal steam. Understanding the thermal regimes of the crust is not only a goal of science, but also a guidebook to finding and evaluating needed energy and mineral resources.

Our present understanding of the regional nature of various thermal regimes is seen in Figure B-1, a map of heat flow in the conterminous United States. The net outward flow of heat across the earth's surface is a measurable quantity that contains information about the thermal regime of the crust. For general discussion it can be related to the processes that generate, transport, and store heat in the earth's crust.

Heat is input to the base of the crust and is enhanced or depleted by processes in the crust. Heat flow at the surface reveals these processes by inference. The processes, in general, are threefold:

1. Contribution of heat from radioactive or rarely some other source in the crust—the heat production component.
2. Heat convected by movement of water, magma, or solid rock—this heat convection component is dependent on heat capacity and vertical velocity of movement.
3. Transient storage or release of heat associated with temperature changes—an example of this transient component is the seasonal climatic effect—heat is stored in near-surface materials in summer and released in winter.

In the more stable, older regions of crust, including large portions of the eastern and central United States, the thermal regime can be accounted for by a simple conduction model involving only the heat input into the base of the crust and the heat production within the crust. Evidently, the mantle heat flow into the base of the crust is uniform over large areas, and the stable regions of higher heat flow are those where the crustal rocks have higher radioactivity. Within stable regions these areas are the most favorable for geothermal development, particularly where crustal rocks with high heat pro-

90

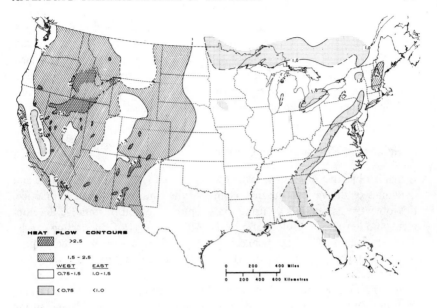

FIGURE B-1   A generalized heat flow contour map of the conterminous United States (Lachenbruch and Sass, 1977).

duction may be buried beneath blanketing sedimentary deposits with low thermal conductivity.

In much of the western United States, where volcanic or tectonic activity has occurred in the last 30 m.y., interpretation of the conductive and convective thermal regimes is much more complicated than in old, stable regions. Drilling can help to clarify these complexities. How has volcanism or tectonism contributed to increased convective heat flow from the mantle into the crust? How have these increased heat inputs been distributed into the higher crust by convecting subsurface waters?

Areally, on the heat flow map, high heat flow anomalies show two distinct patterns: regional highs such as that in northern Nevada, and local high heat flow anomalies such as Yellowstone Park in Wyoming, or The Geysers–Clear Lake area in northern California.

The present explanation of the local anomalies calls for young intrusions of magma, often at less than 10-km depth, with overlying hydrothermal convection systems. Many aspects of the hydrothermal convection systems and their related heat sources are not yet understood. Wide disagreement still exists on the basic explanations for differences between vapor-dominated reservoirs (Larderello, Italy; The Geysers, California) and hot-water reservoirs (Wairakei, New Zealand; Salton Sea, California). Current explanations of the high heat flow regional anomalies are more complex and controversial. Thinning of the lithosphere in regions of extensional tectonics implies mechanical flow of rock toward the surface, in effect enhancing the deep convective heat flow term by upward movement of hot solid rock. Dikes intruded into the crust to accommodate extensional stretching without thinning could also explain the higher regional heat flow. Hot-water convection systems may exist in these regions related to circulating meteoric waters in deep faults, and shallow magmatic intrusions are not necessarily involved.

I. Scientific Rationale for Drilling

A major objective of earth science research is to understand the structure and dynamics of the various thermal regimes of the continental crust. In practice, this objective is realized by using the scientific method of iterative development and testing of geological models. Although part of this testing can be done through surface geochemical, geophysical, and geological techniques, the understanding of thermal processes and structure that can be gained from surface methods is unavoidably limited and in great part inferential. Accordingly, drilling is essential in providing the only means for the direct sampling of fluids and rocks at depth, as well as the only way of directly measuring critical *in situ* physical parameters. In short, drilling is an essential component in the synthesis, refinement, and verification of three-dimensional models of thermal structure and processes in the continental crust.

On the whole, the reliability of geological concepts is inversely proportional to the depths at which they are thought to apply. There is no soundly based understanding of any hydrothermal convection systems below presently drilled depths, roughly 3 km for all types. Convection extends at least to this depth or even deeper in some systems. How are heat and chemical constituents transferred to these convection systems? Are the magma-centered systems supplying some volcanic gases, water, and other dissolved constituents from underlying magma chambers, or are these constituents being supplied by reactions of meteoric water with the rocks of each system? Is the water at the deepest levels of at least some systems exceptionally high in salinity, especially in chloride? If so, are valuable base metals being transported, probably as metal-chloride complexes, and are metal-sulfide ore deposits forming in any of these systems, perhaps with dilution and decrease in temperature? Is a stacked series of convection cells, stratified by salinity and density, characteristic of any of these systems? Is a saline brine essential for major mass and heat transfer at temperatures that are nearly magmatic? Dilute water becomes a supercritical vapor at temperatures above 374°C, regardless of pressure, and is a much poorer solvent of most constituents than water at lower temperatures and brines at higher temperatures. Depending on content and composition of the dissolved salts, brines can persist as liquids up to magmatic temperatures. Thus some brines may be remarkably effective transfer agents for heat and soluble substances at extreme temperatures. For these reasons, brines may be much more effective than dilute water in extending fractures downward into the magma-centered systems. Brines can penetrate into the borders and eventually into the total mass of an originally huge molten magma chamber, thus helping to explain the indicated long lives and great total heat flows of some convection systems (tens to hundreds of thousands of years and even longer, perhaps intermittently).

A comprehensive understanding of the total heat and mass transfer involved in magmatic-hydrothermal systems requires data down to and within the magma body. The scientific challenge of eventually drilling into a magma body is not simply getting there, but the knowledge gained along the way. Although shallow magma bodies are within reach of present drilling depths, it will require revolutionary technology to drill into hostile environments above 400°C. For this reason, the specific targets and subsurface data requirements in this report concentrate on the roots of hydrothermal systems. The results of these initial drill holes will help show the way to deeper slices of knowledge. The panel strongly recommends that research and development in drilling and logging technology be vigorously pursued in order that the ultimate goal of obtaining heat transfer and chemical migration data down to and within magma can be achieved.

II. Subsurface Data Required

The desirable subsurface data for drilling into thermal regimes are described in the *Report of the Workshop on Magma/Hydrothermal Drilling and Instrumentation* (Varnado and Colp, 1978). This workshop was held in Albuquerque, New Mexico, on May 31 to June 2, 1978, as part of the DOE preparation for participation in the Workshop on Continental Drilling for Scientific Purposes. The objectives of the Sandia workshop were to recommend locations for deep drilling into hydrothermal systems, to specify in detail the desired drill hole data, and to assess the state of the art in drilling and instrumentation for such drill holes. The reader is referred to the above document for a detailed discussion of the subsurface data requirements (especially pp. 19–25).

Two important conclusions of the Sandia workshop are worth restating here:

1. Scientific objectives of drilling into magmatic-hydrothermal systems require that representative samples of rocks and formation fluids be obtained and that measurements of temperature and fluid pressure be made. If these minimum scientific objectives cannot be met, there is little point in drilling the hole. If only these data and samples were obtained, they would allow further characterization of other parameters (sonic properties, thermal conductivity, viscosity, density, etc.) under realistic experimental conditions.

2. Several scientific objectives may best be achieved by two closely spaced holes. *In situ* flow properties can be most unambiguously obtained in flow tests utilizing two-hole geometry (e.g., permeability, flow porosity, diffusion porosity, fracture geometry). Horizontal gradients in rock, fluid and geophysical properties, and the correlation of dikes, sills, veins, fracture zones, etc., also require two holes. Furthermore, several technical drilling considerations indicate that one of the most viable methods of penetrating deeply into hydrothermal systems involves drilling an initial reconnaissance hole, in which the sole objective is to drill as deeply as possible.

Although many of the scientific objectives of a continental drilling program will require the drilling of new holes, the value of existing holes and the multiple-use potential of drill holes must be recognized. Some holes drilled primarily to investigate the questions outlined above for thermal regimes may also be used to address other topics. Information about the thermal regimes of the crust can often be obtained from drill holes initiated for other purposes; exploration holes for geothermal and uranium resources are obvious examples. This panel strongly supports a national effort to alert the scientific community to current and proposed drill holes that may have multiple-use potential. A properly designed policy would provide enough lead time to allow the planning of add-on experiments. This practice would maximize the knowledge obtained per drilling dollar.

III. Specific Thermal Regimes Recommended for Deep Drilling

It is clear that the depth and areal distribution of scientific drill hole sampling at present are inadequate to address many critical geothermal problems and that the lack of drill hole information at 3-7 km is a significant impediment to understanding crustal geothermal processes. Accordingly, a carefully designed program to quantify thermal processes in critical deep hydrothermal, magmatic, and other tectonic environments is recommended as follows.

A. Vapor-Dominated Hydrothermal Convection Systems

1. Basic Model
It is now generally accepted that in the drilled parts of vapor-dominated geothermal systems there exists a steam-water reservoir with pressures controlled by steam, although most of the fluid mass is water. The origin of this type of system, rather than the more usual hot-water reservoir, remains controversial. Below the vapor-dominated zone drilled to date, there probably exists a brine-saturated zone, with boiling throughout and with salinity increasing with depth. This deep, water-saturated reservoir has not been iden-

tified in drill holes, and the thicknesses of both the two-phase and the deep brine zones are unknown. The nature of the heat source is presumed to be an underlying young igneous intrusion, but the mechanisms of heat transfer at deep levels and the origin of the vapor-dominated reservoir are essentially unknown. Resolution of these unknowns would be of great practical significance in evaluating longevity and optimum production mode.

2. Target—The Geysers

The Geysers geothermal field in northern California is the largest producer of geothermal electricity in the world (502 MW now; 1124 MW by 1981) and is without rival as the prime candidate for scientific drilling of a vapor-dominated geothermal system. Despite the several hundred drill holes to depths of as much as 3 km, many scientific and practical questions remain, and the size and longevity of the reservoir are unknown. The two-phase reservoir has a temperature of 240°C, but isotopic geothermometer studies suggest fluid temperatures of more than 350°C at depth. The area underlain by commercial steam is at least 70 km^2; the age of the system, from heat flow and heat storage considerations, is probably at least 50,000 years. Target depth for research drilling is 6–7 km.

The Geysers geothermal system is developed in the Franciscan formation, a complex sedimentary assemblage associated with a Jurassic subduction zone. In this formation, ophiolites, cherts, pillow basalts, glaucophane schists, graywacke, and melanges are intermixed, largely with tectonic contacts. The structure at The Geysers consists of high-angle northwest-trending thrust and strike-slip faults that frequently repeat the sedimentary sequence. The vapor-dominated reservoir appears to be bounded on the northeast by the Colloyami strike-slip fault and less certainly on the southwest by the Mercuryville fault. The reservoir is developed in graywacke.

Holes in The Geysers are drilled with mud to 1–1.5 km and then cased and drilled with air into the reservoir. Drilling is stopped when 100–200 tons/hr of steam production is achieved. Some wells have initial wet steam and water production but rapidly dry out to produce slightly superheated steam. Shut-in wellhead pressures are near 32–35 bars, and downhole static temperatures are near 240°–250°C. Pressure gradients are vapostatic. Some wells have perched water tables, but these are almost certainly not the deep water table because temperatures are continuous with those of the steam reservoir. Temperatures will probably be near 250°C at 3.5–4 km and may be above 400°C at 5 km and greater depths.

Pressure drawdown behavior and gas contents within the steam zone suggest that three or four drainage basins may exist, with limited flow between basins. Recharge to the system is from local meteoric water, but natural outflow is limited, in comparison with the size of the system, suggesting very long residence times.

Seismic studies have shown numerous microearthquakes associated with production, but unproduced zones are nearly aseismic. Detailed gravity studies show that production of fluid is from depths near the bottoms of the drill holes and not to a major extent from a deeper (water-filled?) zone. Steam production from local evaporation of water is also supported by isotopic studies.

Young volcanic rocks (100,000–500,000 years, with some as young as 10,000 years) of the Clear Lake volcanic field occur immediately to the northeast of The Geysers. The large positive gravity anomaly associated with the volcanic rocks shows an offshoot under The Geysers, suggesting that the steam field is related to Clear Lake volcanism. However, it is not certain that intrusive or molten rock directly underlies the steam field. Teleseismic P-delay studies suggest that a large volume of partly melted rock underlies the volcanic field at depths of perhaps 5 km, roughly coincident with the gravity anomaly. One or more hot-water geothermal systems underlie the Clear Lake volcanic field, but there are few drill holes, and the temperatures and characteristics of these hot-water systems are poorly known.

3. Major Questions

Do the hypothesized deep brines actually exist, and what is their possible role as ore fluids? What is the relative importance of these deep brines and of dispersed pore water in supplying steam to wells? What is the relation of vapor-dominated systems to high-level magmatic intrusions? What are the conditions that lead to a vapor-dominated system rather than a hot-water system? In the hypothesized deep brine, do the nearly constant temperatures of the reservoir rise with depth, first along a boiling point curve controlled by the increasing pressure of brine and then by a linear conductive gradient? Or does the brine convection system extend downward into the original magma chamber?

4. Critical Data Needed in Order of Priority

These are as follows:

- Temperature as a function of depth
- Fluid pressure as a function of depth
- Fluid composition (liquid and gas; chemical and isotopic)
- Cuttings and selected cores of rocks
- Other geophysical logs, especially those bearing on porosity and permeability.

B. Hot-Water Hydrothermal Convection Systems Driven by High-Level Igneous Intrusions

1. Basic Model

Hot-water convection systems driven by high-level igneous intrusions are the most common type of geothermal systems developed to date for electricity and are widely distributed in volcanic belts around the world. In general, drilling has defined single-phase (water) convective reservoirs at temperatures up to 360°C, presumably maintained at high temperature by underlying igneous intrusions. Although subsurface young intrusive rock has been encountered in a few systems (e.g., Krafla in Iceland and Salton Sea in California), the nature of heat transfer from the intrusion and the character of the deep zones of hydrothermal convection remain obscure. These zones presumably are appropriate for the formation of hydrothermal ore deposits, but the chemical and physical conditions that actually control transport and deposition are poorly known.

2. Targets

In many respects, the Yellowstone caldera is the outstanding geothermal target in the United States for deep research drilling. Geological, geochemical, and geophysical data indicate that the Yellowstone Plateau volcanic field overlies a group of plutonic bodies of batholithic size emplaced over a period of 2 m.y. The older parts of the plutonic system are probably solidified but still hot. The youngest of these very large magma bodies has been emplaced during the last 1.2 m.y. Two higher-level parts of this magma chamber formed a pair of adjoining ring-fracture zones, through which a climactic pyroclastic eruption of more than 1,000 km^3 of molten matter occurred 600,000 years ago. The roof of the magma chamber collapsed along the two ring-fracture zones to form a 40 by 70 km caldera. Postcaldera rhyolitic volcanism has continued intermittently for the past 600,000 years. An episode of intracaldera doming and three major sequences of rhyolitic lava eruptions have also occurred during the past 150,000 years, the latest about 70,000 years ago. Geophysical evidence strongly supports the existence of molten magma in the present volcanic system at depths near 6–7 km. The distribution and ages of vents for the younger volcanic eruptions clearly indicate episodic resupply of magma to shallow parts of the system from deeper levels. Heat must be supplied continually to the rhyolitic magma chamber from deeper basaltic magma, represented at the surface by basaltic lavas around the caldera margins, whose eruption spanned the entire history of the rhyolitic field.

A massive hydrothermal convection system exists above the now partly solidified Yellowstone magma chamber. Deep recharge to the system may be principally from the north, with shallow recharge and dilution from local sources. Discharge occurs in many areas, both as hot-water and as vapor-dominated systems, controlled mainly by two ring-fracture zones. The most spectacular thermal features are the geysers. Temperatures as high as 237°C were measured in research holes drilled in 1967-1968. Isotopic geothermometers and mixing models indicate subsurface temperatures generally as high as 360°C, with salinities normally increasing downward because of decreasing local recharge and dilution. A major challenging question is: Are the observed modestly saline 360°C waters the most saline of the system or do deeper brine levels exist, perhaps gravity-stratified but higher in temperature, and "unseen" by any of our present techniques?

However, Yellowstone is our first and most widely known national park, with natural thermal activity that is more extensive and diverse than anywhere else in the world. Geothermal exploration and exploitation have already destroyed some of the other major geyser areas of the world (Wairakei and Tauhara, New Zealand; El Tatio, Chile; and Beowawe, Nevada). The disturbing effects can be far-reaching. For example, exploitation at Wairakei has drawn off fluids and pressure from the seemingly independent Tauhara-Taupo system centered 6 km to the southeast and has destroyed the geysers at The Spa on the border of this less well-known system. Suitably located and carefully monitored deep research drilling in Yellowstone, accompanied by *minimal discharge of fluids*, could avoid such destruction. However, the environmental disturbance necessary for drilling on this scale, and the implied threat of possible future exploitation, may be unacceptable to the National Park Service, to the scientific community, and to the concerned public—all of whom are interested in preserving our unique heritage. Any proposal for deep drilling in Yellowstone must emphasize the research aspects and must demonstrate committed and strongly convincing opposition to all future possibilities of energy production or other economic use.

In view of the potential difficulties of drilling in Yellowstone, the panel does not recommend deep research drilling in Yellowstone at this time, but instead suggests two alternative geothermal systems.

a. Salton Sea geothermal system, California. The Salton Sea geothermal system is located in the Imperial Valley, where the East Pacific spreading system passes under the Colorado River delta and then is offset hundreds of kilometers to the northwest along the San Andreas fault. The geothermal system is in Pliocene and Quaternary delta sediments and is associated with the young (approximately 16,000 yr) rhyolite domes and abundant subsurface intrusive rocks of silicic and mafic composition. The geothermal system and the igneous rocks appear to be related to a short spreading segment between the Brawley and the San Andreas transform faults.

Fluids from the Salton Sea geothermal system contain as much as 250,000 mg/l dissolved solids, primarily chloride, sodium, potassium, and calcium, with low sulfur and a host of heavy metals (iron, manganese, zinc, copper, silver, etc.). Temperatures are at least as high as 350°C in the reservoir at 3000–6000 ft. The host delta sediments have been hydrothermally metamorphosed to the greenschist facies.

Although the Salton Sea geothermal system has been known and drilled from 1 to 2.5 km depth since the early 1960's, utilization to date has been hindered by the very high salinity and temperature, with consequent serious corrosion, scaling, and brine disposal problems. Department of Energy experimental work continues, and renewed industry attention is now evident.

The rhyolite domes, the intrusive rock encountered by drill holes, the large positive magnetic anomaly, and the tectonic setting clearly indicate the presence of a young batholith at shallow depths (5 km?) in the late Cenozoic deltaic sediments. The youth of the igneous rocks, the intensity of the hydrothermal system, and the tectonic setting strongly suggest that much of the igneous material at or near the bottom of the sedimentary section (6–7 km) may be molten. This geothermal system is also of major interest for its ore constituents and for its location on the southern limits of the San Andreas fault.

b. Valles caldera, New Mexico. The Jemez Mountains of northern New Mexico have been the focus of intense volcanism throughout much of the Pliocene and Quaternary. A period of basaltic, andesitic, and dacitic volcanism 2.5 m.y. ago was followed by two major ash flow eruptions accompanied by caldera subsidence. The lower member of the Bandelier tuff erupted from the Toledo caldera. The upper member erupted from the Valles caldera, 24 km in diameter, 1.1 m.y. ago. Caldera collapse associated with this last ash flow eruption was followed by reinjection of magma that caused resurgence of the caldera floor and culminated in a ring of rhyolite domes along the caldera margin; the youngest dome is only 100,000 years old. Within the caldera a major high-temperature hot-water geothermal system is probably capped by a small vapor-dominated system; reservoir temperatures exceed 290°C. The area comprises private land (the Baca location) leased to Union Oil Company, which has drilled about 25 exploration wells. Although virtually all the exploration and reservoir data are proprietary, the Valles caldera is clearly a major commercially exploitable geothermal resource. In early July 1978, agreement in principle was reached between the Department of Energy, Union Oil, and New Mexico Power and Light to construct a 50-MW demonstration plant. In addition, the Los Alamos Scientific Laboratory hot, dry rock experiment is located in Precambrian gneiss and amphibolite just west of the Valles caldera.

The recency of volcanism, the volcanic history, and the presence of a large, high-temperature hydrothermal convection system suggest that still-molten magma underlies the Valles caldera. Published geophysical surveys are incomplete, and the depth to magma is uncertain, but perhaps 5–8 km is reasonable. Salinity of the deep reservoirs is approximately 6000 mg per liter of total dissolved solids.

In summary, the hot-water hydrothermal convection system in the Valles caldera is representative of hot-water geothermal systems associated with young, high-level silicic magmas. Release and evaluation of industry data coupled with systematic geophysical surveys should allow siting of a deep hole to explore the roots of the hydrothermal system and thus elucidate the processes by which the magmatic heat is transferred to the hydrothermal convection system. A deep drill hole will also enhance our understanding of large, cooling magma bodies and the structure of calderas.

3. Major Questions

What is the nature of heat transfer from magma to convecting hydrous fluids? Do deep brines have a role in maintaining overlying high-temperature hot-water systems and in depositing ores? What are the factors that control the balance between fracturing and hydrothermal deposition and thus the longevity and periodicity of hydrothermal convection systems? What are the roles of magmatic, connate, metamorphic, and meteoric waters in supplying ore metals?

4. Critical Data Needed in Order of Priority

These are as follows:

- Temperature as a function of depth
- Fluid compositions (water and gas; chemical and isotopic)
- Cuttings and selected cores of rocks
- Fluid pressure as a function of depth
- Other geophysical logs, especially those bearing on porosity and permeability.

C. Hot-Water Geothermal Systems Driven by Regional Conductive Heat Flow

1. Basic Model

An area of perhaps several tens of thousands of square kilometers in Nevada and Idaho is characterized by heat flow generally in the range of 2.5 HFU, or nearly double the world-wide average. This area also has the greatest abundance of thermal springs (and spring systems with temperatures greater than 150°C) of any large area in the United States. Silicic volcanism younger than 10 m.y. is generally lacking, and the few young volcanic rocks are basalt that show no convincing relations to the hot springs. Hence it seems likely that the abundant hot springs are due to regional high heat flow, rather than shallow igneous intrusions, and to deep penetration of meteoric water along the normal faults of this extensional province. The regional extension may be accompanied by thinning of the crust, basaltic intrusion into the lower crust, and relatively shallow regional metamorphism. Consequently, this region may contain more hot, dry rock at higher temperatures and shallower depths than any other area of comparable size in the United States.

2. Target

No specific target for deep drilling can be recommended at present because of inadequate regional geophysical characterization. Expanded regional heat flow surveys, hydrological analysis, and active seismic surveys should be integrated with geological, magnetic, and gravity data to define specific candidate targets for deep drilling.

3. Major Questions

What are the tectonic controls of hydrothermal circulation and temperature distribution? Are longevity and reservoir capacity of these hydrothermal systems significantly different from those of systems driven by high-level igneous intrusions? What is the nature of the deep thermal and hydrological regimes? What is the nature of the deep thermal transfer processes? What role does magma play in sustaining these regional heat flows? Does the lower salinity of thermal waters in northern Nevada reflect the low input of volcanic fluids to the region?

D. Thermal and Chemical Processes in Active Volcanic Regions

Conceptual models of active volcanic systems can also be tested by drilling. Basic scientific questions include: What are the relations of individual volcanic systems to larger intrusive systems at depth? How does magma interact chemically and thermally with its wall rocks, and what volatile and soluble constituents does it contribute to hydrothermal systems? May high regional and local heat fluxes at depth be "screened" by shallow hydrological effects? How extensive and important is meteoric water circulation in the evolution of volcanic activity and associated hydrothermal phenomena?

The Hawaiian, Cascade, and Alaskan volcanoes all provide opportunities for specific drilling targets. For example, a scientific and technological program of drilling into ponded magma in Kilauea Iki lava lake is presently under way. This lava lake shows several similarities to magmatic-hydrothermal systems:

1. The thermal regime is driven by cooling and crystallization of magma.

2. The heat transfer between the magma and its surroundings is by conduction and by convection of aqueous fluids in fractured host rocks.

3. The aqueous fluids comprise a two-phase (liquid-vapor) circulation system with a boiling zone.

4. The fractured crust permits circulation of meteoric water down to hot rock immediately over the melt.

5. The basaltic rocks above the melt are undergoing low-temperature ($\leqslant 100°C$) hydrothermal alteration.

Although it is by no means a perfect analog, Kilauea Iki lava lake is similar in some respects to large-scale magmatic-hydrothermal systems. Hence the panel recommends that Kilauea Iki be used as a field laboratory for preliminary investigation of many of the scientific and engineering problems involved in drilling larger and deeper systems.

Another example is provided by the active and dormant stratovolcanoes of the earth's subduction zones. The Cascade Range, the Alaskan Peninsula, and the Aleutian Islands are characterized by chains of andesitic stratovolcanoes. The complex internal structure and shallow magma chambers within and below this type of volcano are not well understood. A drilling program on such a volcano would be necessary to understand its internal framework and hydrothermal systems. Several of the Cascade volcanoes have erupted during the last century, and current microearthquake activity indicates that some still have dynamic systems. Specific drilling targets might be Mt. Baker or Mt. St. Helens in Washington or Mt. Shasta in California. The diversity and number of stratovolcanoes in the United States preclude an intelligent drilling choice at this time.

E. Thermal Processes in Stable Regions

The conceptual model of conductive heat transfer into the base of the crust and additional generation of heat within the crust from radioactive decay has been well established for stable crustal regions. However, the distribution of radioactive (heat-generating) elements with depth in the crust needs additional documentation, and the role of hydrological disturbance in many areas is imperfectly understood.

Mantle heat flow appears to be uniform over large areas of the eastern United States, with the areas of somewhat elevated surface heat flow being those where the crustal rocks have higher radioactivity than normal. Although heat flow in the eastern United States is only low to moderate, the region contains many population centers and thus has a high energy demand. It is thus important to understand the thermal regimes of this stable region, both because of the potential for supplying thermal energy for direct use and because such an understanding will help us to develop models of more complicated regions.

It does not seem appropriate to recommend a specific site or sites for deep drilling solely for thermal and heat generation measurements. However, any drilling for scientific purposes in basement rocks should make provision for temperature, thermal conductivity, heat generation, and hydrological measurements. Provision should be made for similar measurements in holes drilled by federal and other agencies under mission programs, and a clearinghouse mechanism should be set up to ensure that opportunities for thermal measurements are not lost through failures in communication. Furthermore, systematic delineation of regional heat flow provinces and analysis of hydrological anomalies should continue, using relatively shallow (< 300 m) holes, both holes of opportunity and holes drilled specifically for thermal and hydrological studies.

IV. Concluding Statement

Drilling is an expensive scientific tool. One hole to 6 or 7 km in each of three thermal regime targets and the sampling and logging to attain the needed scientific data is a $30 million program. The scientific effort related to this program, including site selection, laboratory examination of samples, and interpretation, requires a comparable magnitude of funding. This clearly illustrates the importance of optimum site selection, scientific monitoring during the drilling, and adequate examination and storage of samples and logs.

Site selection involves integration and interpretation of the scientific data from at least five sources: (1) geological and geochemical investigations of the surface, and inferences on the subsurface structure that best explain the surface data; (2) geological and geochemical investigations of cooled and exhumed magmatic-hydrothermal systems; (3) geophysical surveys that yield data on the subsurface conditions—heat flow, seismic, electrical, magnetic, and gravity; (4) shallow- and intermediate-depth holes drilled into hydrothermal systems for geothermal resource exploration or development, and a few shallow holes drilled for scientific purposes into hydrothermal systems and lava lakes; and (5) theoretical modeling of magmatic intrusions cooled by interaction and convective transport of groundwater. Additional constraints to site selection include access, topography, and environmental considerations. These are not trivial, but except for Yellowstone, they were not considered by this panel. Table B-1 is a summary of prime targets.

Concerned scientists should be on site during the drilling. They should have sufficient authority, in conjunction with the drilling engineers, that opportunities for obtaining significant scientific data can be identified and acted upon during the drilling process.

Samples and logging data should be available to serious scientific investigators regardless of affiliation. A specific organization should take responsibility for the allocation and storage of the precious samples and logs.

TABLE B-1 Summary of Drilling Targets Related to Thermal Regions

	The Geysers Geothermal Area, California	Valles Caldera New Mexico	Salton Sea, California	Synthesis of the Three Prime Targets	Normal Heat Flow in the Eastern United States and Midcontinent	High Regional Heat Flow
Scientific significance	Test model that vapor-dominated reservoir is underlain by hot brines. Evaluate nature of brine formation and its character at high T and P.	Nature of large silicic magma bodies and caldera structure. Characterization of modestly saline wet hydrothermal system of probable economic significance.	What is the function of brine in heat transfer? Drill into environment where metamorphism is occurring.	Heat and mass transfer below the explored systems, geothermal and/or igneous. Evolution of igneous systems.	Distribution of heat sources in crust. Dominantly radioactive K, U, Th. State of stress. Test models of radiogenic heat production: constant, exponential or other distribution of radioactive nuclides with depth.	Origin of anomalous heat: hydrological or tectono-magmatic?
Potential practical importance	Estimation of longevity of geothermal power production. Knowledge of mass balance of water important to optimum production procedures. Possible direct observation of ore deposition process.	Explore relationships at depth for models of thermal state, ore deposition and geothermal resource base. Determine hydrothermal reservoir properties and assess likely longevity of hot-water system.	Technology development (major problem in using brines for geothermal energy). Nature of reservoir and permeability of systems. Observe the deposition of ore deposits if already present.	In situ studies of base metal minerals being deposited. Nature of permeability, in different geothermal and igneous systems. Fluid mass balance.	"Low-grade" geothermal resource for space heating, cooling and industrial processes. Deep waste isolation. Possible information related to earthquake hazards.	Largest potential regional geothermal resource (hot, dry rock and hydrothermal).
Status of Data	Geological maps (1 in. = 1 mile). Geochemistry (water, petrology work).	Geological maps (1 in. = 1 mile). Geochemistry (partial data on rocks and water).	Geological maps (surface and subsurface). Geochemistry (adequate, except	Deep geophysical surveys inadequate. Encourage COCORP to generate	Regional heat flow, geohydrological, and geological maps.	Regional reconnaissance heat flow and geological maps.

104

	Geophysical surveys (deep geophysical surveys inadequate).	Geophysical surveys (partial). Volcanic petrology (partial).	volatiles). Geophysical surveys adequate, except for deep measurements.	seismic reflection profiles of these targets.		
Drilling Experience	200 wells, up to 3-km depths Producing geothermal field.	25 wells, to about 2-km depth in caldera. 3 wells, up to 3-km depth, outside edge of caldera.	15 wells to 2.5-km depth. Modest production from flanks of the field.		Numerous wells (Deepest well >10,000 ft), in granitic rocks.	25 wells up to 3-km depth at thermal spring systems related to the regional anomaly.
Problems	Drill *below* present drillable depths. Differential pressure across casing.	Preparing to produce. No *special* problems.	Extreme temperatures and salinities.	At depths of 3.3 km, temperature is biggest problem. Coring at >350°C. Instrumentation at high temperatures.	None foreseen.	Target depth is 7 km; no problems unless $T > 350°$ C.
Existing models of systems	Hydrogeological, thermal, tectonic, geophysical.	Geological, caldera evolution, hydrogeological, thermal.	Hydrogeological model. Geophysical model. Tectonic model. Thermal model.	Numerical models of hot-water convection systems and cooling magma bodies based on surface and drill hole data, size, age, composition, for all three systems.	Models of conduction and fluid motion—which is correct? or are both correct?	Tectonomagmatic model. Hydrogeological model.
Selected Recent Reference	*Second U.N. Symposium on Geothermal Resources,* 1976.	Kolstad and McGetchin, 1978.	Robinson *et al.,* 1976.		Lachenbruch and Sass, 1977.	Lachenbruch and Sass, 1977.

Appendix C
Mineral Resources

C MINERAL RESOURCES

I. General Objectives

A. Introduction

After six millenia of mineral production, we still do not understand how and why many kinds of mineral deposits form. As a result, our ability to find additional deposits is restricted. This is especially so for metallic mineral deposits, and it is from this group that we will benefit most by having a systematic and focused program of continental drilling.

The initial benefits of such a drilling program will be more scientific than economic. The major goal is to gain a better understanding of the special geochemical circumstances that lead to the formation of mineral deposits. This translates to a program of understanding the processes that generate, transport, and concentrate metals in the earth's crust. The program must therefore be aimed at the studies that arise from drilling the holes, not at the drill holes themselves. This means that a successful drilling program must provide for the costs of scientific studies as well as the costs of the holes.

While drilling is common in and around deposits that have been mined, the holes have rarely been drilled solely for scientific purposes. Rather, they have been located to answer practical questions of mining. Mineral deposits and deposit-forming processes have usually been studied within the mineral volumes of the deposits. Yet the deposit-forming processes involve much greater volumes of the crust than the deposits themselves. For example, the mined and explored volume of a porphyry copper deposit may be as large as 2 km³, but the rock volume involved in the formative processes may be as large as 100–200 km³.

Although some deep drill holes are needed for the proposed study, many vital scientific questions may be answered by relatively shallow drilling, to about 5000 ft ($\approx$ 1500 m), provided the holes are properly located. Furthermore, most of the holes proposed for a program designed to study mineral deposits will be available and useful for other purposes, such as downhole geophysical experiments. The essential thing is to provide, for the first time, an opportunity to study mineral-depositing processes in three dimensions. The result will be a major advance in our understanding of how mineral deposits form; thereby we may anticipate a revolution in our approach to seeking and finding new deposits. *Three-dimensional mapping of the crust, focused in large part on important mineralized areas, is the next great advance needed by the geological sciences.*

B. The Program

A program of continental drilling will contribute to three general areas related to mineral deposits:

- *Studying the fundamental geological structure and composition of the continental crust.* Mineral deposits are rare geological phenomena. The nature of mineral deposits formed by natural chemical processes and the frequency with which the processes have operated have varied through time and with position in the crust. Different kinds of deposits are, as a result, associated with specific regions of the present crust. Yet we are still unable to say whether observed distributions are strictly a result of compositional variations in the crust or whether localizing processes may produce certain deposits regardless of crustal composition. Drilling may provide the samples needed to resolve the question.

  Additional important questions may be answered by suitable drilling programs. For example, we still do not know with certainty whether the amount of a given metal available for mining increases geometrically with declining ore grade, even though that has been widely assumed. Nor are we sure of the distribution of mineral deposits with depth. We do know that most mineral deposits form within 5–6 km of the earth's surface, and we surmise that the frequency of deposits in the present crust decreases with depth. However, large-scale movements in the crust during geological time may well have buried once shallow deposits deep in the crust. Until more is known of the structures of the crust and the rock types it contains, the question of depth and frequency of deposits remains conjectural.
- *Observation of active systems of the kind that have formed mineral deposits in the past.* A large number of the scientific problems may be studied in active hydrothermal (hot-water) systems. However, only the high-salinity hydrothermal systems are likely to be major transporters of metals. Most hydrothermal systems that have been studied so far are heated by magmatic heat and are dilute solutions. However, it is probable that some vapor-dominated geothermal systems are underlain by brines of the kind that form mineral deposits, and we propose to test this important hypothesis.
- *Studying past mineralization.* Because mineral deposits are rare geological features, it is unlikely that we can find and study a present-day hydrothermal system in the act of forming a mineral deposit. We must therefore

explore some successful fossil hydrothermal systems, to complete the link between them and mineral deposits.

C. Summary Estimated Cost of the Program

The mineral deposits program may be evaluated as to cost for both drilling and science programs. To estimate drilling costs, we used those supplied by the Department of Energy. To estimate science costs, we have used our best experience. Cost estimates are for university studies. The results are summarized below.

Project	Drilling Cost, Thousands of Dollars	Science Cost, Thousands of Dollars
Precambrian iron basins	750–1,220	600
Volcanogenic sulfide deposits	1,720	1,000
Alkaline igneous complexes	1,720	1,000
Chemistry of the crust	4,140	4,700
Magmatic-geothermal systems	7,800	4,600
Sedimentary basin water	–	1,000
Microfracture studies	–	2,500
Mineral deposits	14,400	10,000

D. Site Priorities

The Panel on Mineral Deposits strongly recommends that most targets be considered as programs, rather than requests for single drill holes. The top four projects discussed in part II of this appendix are as follows:

- A deep hole into and through the Salton Sea geothermal brine pool.
- A program of shallow and deep drilling below the epithermal, gold-silver deposit at Tonopah, Nevada.
- A program of deep drilling, extending the extensive shallow drilling by Arco-Anaconda, beneath the mineralized zone of Butte, Montana.
- A program of shallow and deep drilling in the Precambrian Animikie Basin, Minnesota.

II. Specific Objectives

A. Structure of the Crust

There are few imaginable features that deep drilling might encounter that would not be of interest to those who study mineral deposits. Knowledge of the continental distribution of rock types and ages would inevitably improve our understanding of continental evolution, thereby aiding both mineral exploration and our understanding of some of the basic principles of mineral deposit genesis. Certainly the mineral industry would benefit from a knowledge of the distribution of the Archean rocks within the continental mass. Likewise, the depths, structures, record of evolution, and composition of the basement rocks of deep basins (such as those of the Basin and Range or the Triassic rift of eastern North America), the Keweenawan rift system of the midcontinent, the Michigan Basin, the Illinois Basin, and the Mississippi Embayment are of considerable interest with regard to their possible role in the formation of the ore deposits that flank them, as well as the possible existence of new types of ore within them.

Although there are many intriguing possibilities for studying specific continental structures and properties, from the standpoint of mineral resources, we suggest the following structures as the most important, but at the same time recognize that a Continental Scientific Drilling Program might solve a great many additional problems.

1. Precambrian Iron-Bearing Basins

The origin of the widespread Precambrian iron formations of the so-called "Lake Superior type" has been the subject of research and intense speculation for many years. Of particular interest are the middle Proterozoic iron formations of the Animikie Basin, which are the largest in the United States. Iron-bearing strata are exposed around the fringes of the basin, where they have been explored and exploited for many years. As a result, a large body of data bearing on their distribution, internal stratigraphy, mineralogy, and composition is available. The extensive data base has led to a number of models that attempt to explain the origin of iron formations in general; in fact, the Animikie Basin has long been considered a prototype for iron formation deposition. The importance of the origin of the Animikie Basin to the general understanding of the evolution of iron formations has been recently reemphasized in connection with the IUGS-UNESCO International Geological Correlations Program, Project 132 (Trendall, 1978).

The observation that the general stratigraphic succession on the north side of this basin (Mesabi and Gunflint ranges) is remarkably similar to that (Gogebic range) on the south side led many early geologists to postulate that the iron-rich strata were a widespread blanket deposit, which was later deformed to its present synclinal configuration by subsequent tectonic processes. Later studies, particularly of clastic strata both below and above the main iron formations, have suggested that the iron-rich strata were deposited in an intracratonic basin bounded and underlain by Archean rocks of diverse lithic character and age. As a consequence, the iron-bearing units are now considered to be shallow-water, strand-line deposits that, most likely, were never physically connected. Similarly, there is considerable controversy about the origin of the iron and silica now found in the iron formations. Although some geologists have suggested that these constituents were derived by the weathering and leaching of country rocks around the basin, it now seems more likely that they were derived from volcanic rocks within the basin itself.

Although all of these ideas are plausible, none may be considered proved, because critical data from the axial portion of the basin are totally lacking. Further, it is unlikely that the private sector will be able to obtain these data in the near future. Therefore a federally supported scientific program of deep drilling may be the only means for obtaining such information. The program would lead directly to much-needed stratigraphic data bearing on the vertical and lateral succession of rock types, on the distribution of textural and chemical facies within the iron formations themselves, on the possible presence of manganese, iron, and phosphorus deposited in potentially economic concentrations, and on the nature of the substrata upon which the sedimentary rocks were deposited. The latter information would provide constraints on various models that have been proposed to explain the tectonic evolution of the Animikie Basin.

Most importantly, however, subsequent scientific study of the materials acquired by drilling would provide the petrographic, mineralogical, and chemical data necessary for a better understanding of the exact sedimentary conditions and environments that caused large-scale deposition of these iron formations, and to the diagenetic and/or metamorphic processes that subsequently affected them. Clearly, when these data are integrated with those from other iron basins throughout the world, they will lead to a more comprehensive understanding of this enigmatic class of mineral deposits.

Although various kinds of geophysical studies, including both shallow- and deep-seismic profiling, would be required, currently available geophysical data suggest that this program could be carried out by drilling four to six wireline-cored NX (3.5 in., or 88.9 mm) diameter holes, sited along a line perpendicular to the strike of the Mesabi Range in northern and east central Minnesota. It is anticipated that the shallowest hole would be sited so as to intersect about 2,000 ft (700 m) of strata, whereas the deepest hole in the axial portion of the basin would intersect 10,000–12,000 ft (3,000–4,000 m) of strata before penetrating basement rocks. This program would yield approximately 24,000–42,000 ft (8,000–14,000 m) of core for an estimated cost of $746,000–$1,219,000, depending on the specific number of drilling sites. Additional funds for scientific studies of the holes and cores, over a 3- to 5-year period, would amount to approximately $600,000.

2. Distribution of Precambrian Volcanogenic Sulfide Deposits

The belt of Precambrian volcanic and sedimentary rocks extending across northern Wisconsin contains several massive sulfide deposits that are rich in copper and zinc. Although the belt has characteristics resembling those of Archean volcano-sedimentary belts in the Superior province of the Canadian Shield, its nature is still inadequately known because outcrops are restricted. Several cycles of volcanism and sedimentation appear to be represented, but at present the lateral extent, stratigraphic succession, and structure of the rocks are undefined. The mineral deposits are integral components of the volcano-sedimentary system (or systems) present in the belt, but because the system cannot be defined from surface mapping or from the drilling that has been done in very limited portions of the belt, efficient exploration of the belt is impossible.

The Precambrian belt passes both westward and eastward under Paleozoic cover and may thus be of far greater extent along strike than indicated by surface mapping. If deposits are ever to be found beneath the Paleozoic cover, the nature of the Precambrian belt must be better understood. Drilling is the obvious means by which the nature and structure of the belt may be determined and the volcano-sedimentary system or systems sampled.

We propose a two-phase program: the design of the second phase is to be based on the results of the first phase. In the first phase of shallow drilling, four lines of 20 holes each across the belt should be drilled to depths of about 500 ft (150 m). Spacing of holes along each line would be based on what is currently known about the bedrock geology; lines should be spaced about 15–20 mi (25–39 km) apart. All holes should be NX (3.5 in., or 88.9 mm) and wireline cored. Each hole should be inclined so as to achieve maximum intersection of the stratigraphic succession. A reentry capability should be maintained on all holes. Cores should be sampled and analyzed by standard mineralogical, petrographic, and chemical methods. Samples should be taken for measurement of physical properties, such as magnetic susceptibility, density, and resistivity, as an important objective of drilling is to provide a sounder basis for interpretation of geophysical surveys. Cost of Phase 1 of the drilling program is estimated at $860,000 for 40,000 ft (13,000 m) of drilling. The additional cost of scientific investigations is estimated at $500,000.

Phase 2 of the drilling program, which will probably cost at least as much as Phase 1, would involve deeper holes that are planned only after Phase 1 results have been evaluated. It is possible that some of the Phase 1 holes might be reentered and deepened and that add-on experiments might be possible by deepening exploration holes drilled by mining companies.

3. Structures and Mineral Potential of Alkaline Igneous Complexes

Alkaline igneous complexes are the world's major sources of niobium and rare-earth elements, important sources of phosphate and copper, and actual or potential sources of uranium, iron-titanium minerals, and beryllium. The most productive and best understood complexes lie outside the United States. Such complexes do occur in the United States, but they are incompletely investigated, not well understood, and may contain significant mineral deposits at depth. A better understanding of them is highly desirable, not only for estimating mineral resources, but also for understanding the structures and origin of alkaline complexes.

The alkaline complex at Magnet Cove, Arkansas, is the best known in the United States, and its surface geology has been mapped in detail. Nothing is known, however, of its structure and composition at depth. The prime objective of drilling would be to determine changes in the nature and configuration of the rock units below the surface—particularly to determine whether larger bodies of carbonatite occur at depth.

Phase 1 of this program should consist initially of eight holes spaced along a selected diameter of the complex, each hole NX (3.5 in., or 88.9 mm) wireline cored, and drilled 5000 ft ($\approx$ 1500 m) deep. Each hole should be inclined 60° toward the center of the complex. Deeper drilling in phase 2 should depend on results of the initial phase. The cost of drilling eight holes is estimated at \$860,000. Cores should be sampled and analyzed for minerals present and for major, minor, and trace elements, by standard methods. Cost of logging and the scientific investigation of cores is estimated at \$500,000. Phase 2 costs will be at least as large as those of phase 1. Add-on experiments are unlikely because no relevant drilling is currently under way.

The detailed surface mapping of the complex is available. It should be supplemented by gravity, aeromagnetic, and radiometric surveys.

B. Chemistry of the Crust

Mineral deposits usually represent the extreme limits of natural fractionation processes that affect the entire crust. For some metals, we rely on deposits that are 10,000–100,000 times more concentrated than the average crustal abundance; the availability of these materials to society depends initially on the amount of fractionated material available. The exploration for new metal deposits may become more effective if a genetic understanding of the deposit-forming processes is available. We need both an observational program to furnish data on the fractionation of metals in the crust and a research effort to develop an understanding of the processes of concentration.

1. Grade-Tonnage Curves

Skinner (1976) has suggested that the geochemical contrast between atomic substitution of trace metals in rock-forming minerals and the existence of ore minerals in which the same metals are essential components leads to a bimodal distribution on a grade-versus-tonnage plot. Because the consequences to society of depleting the high-grade peak of this bimodal curve and finding no lower-grade ores is so enormous, the testing of Skinner's hypothesis is of major importance.

However, the volume of data needed for such a test is extremely large. Only by designing a weighted-sampling scheme can the data problem be reduced to manageable proportions. A suitable sampling program must examine areas with high metal concentrations most intensively. An automated scanner capable of rapidly examining unprocessed cores offers the best hope of providing sufficient data. Experience with an X-ray-fluorescence core scanner (Deffeyes, 1969) indicates that metals like lead, zinc, manganese, and copper may be determined at levels below the average crustal abundance at scan rates above 1 mm/s along the core. For silver, tungsten, mercury, antimony, cobalt, chromium, and nickel, the X-ray-fluorescence technique would have detection limits about 2 orders of magnitude lower than the present economic limits of mining. A wide range of modern equipment and methods should be examined for possible use in a rapid analytical method that could be incorporated into core scanners.

Drill cores are essential in order to avoid the effects of near-surface weathering. Fortunately, existing cores may be analyzed to provide an initial data bank, but access to cores from a well-designed continental drilling program will be essential to complete the data bank.

2. Metallogenic Provinces

Although economic geologists have usefully employed the concept of metallogenic provinces for years, we still do not know whether the crust in these provinces is any richer in certain metals than in other areas. The possibility exists that metal-concentrating processes have mobilized trace amounts of metal from large amounts of rock and created valuable metal deposits in one area, but not in others.

Areas selected for initial study would be chosen because of known concentrations of metal deposits (such as southeast Arizona), for clear relations to plate tectonic settings (Cenozoic rocks of the Great Basin and the volcanic region of southern Alaska) and for age differences (Superior province). In addition, core from those holes that have no relevance for metal exploration (such as oil exploration and Department of Defense holes) should be examined to provide baseline data.

The highly mineralized region of southeastern Arizona is such an important and remarkable metallogenic province that we propose a specific drilling program to examine the geochemistry of this region. The 1.6- to 1.7-b.y.-old orogenic belt in the region consists of a eugeosynclinal volcanogenic-graywacke suite, the Precambrian Pinal schist, that is intruded by a series of granitic plutons. Since these lithologic units underlie the entire copper-rich province, answers to the following questions are of great interest:

- Is this portion of the orogenic belt geochemically anomalous, in relation to portions of the same belt that underlie barren areas well outside the copper province?
- Is there a unique geological terrane beneath the exposed Precambrian rocks that is the cause of the copper-rich province?
- Are there sedimentary facies changes, structural boundaries, or transitions in metamorphic grade that distinguish the copper province from the adjacent barren areas?
- Would the additional structural, petrological, and geochemical information obtained from drilling within the copper province help understand the distribution of individual deposits within the province?

To derive the answers, we propose a drilling program involving four holes about 5,000 ft ($\approx$1,500 m) deep. Information from these holes plus the available regional geological and geophysical information will allow the siting of a single 20,000- to 25,000-ft ($\approx$ 7,000–8,000 m) hole to sample the Precambrian basement. Complete geochemical characterization of core from this hole and comparison with cores from other "normal" areas of the crust should allow us to determine whether anomalously high concentrations of copper or evidence of widespread leaching of copper is characteristic of the deep basement. For comparison, it would be useful to have data from another 10,000-ft ($\approx$ 3,000 m) hole in the same 1.6- to 1.7-b.y.-old terrane unrelated to the copper-rich province—perhaps from the Zuni Mountains uplift in north central New Mexico.

Studies of the drill core should be integrated with a systematic outcrop sampling program in this region. However, the great advantage of studying deep drill core is that we are much more likely to be able to evaluate rocks that have suffered neither the effects of surface weathering nor the effects of shallow low-salinity hydrothermal activity. Inasmuch as we are looking for leaching effects, it is essential that we obtain the deepest possible samples from beneath the zones of Mesozoic and Cenozoic hydrothermal activity.

3. Sedimentary Rocks

At present, more than half of our metals already come from sedimentary rocks; this proportion is increasing with time. There are several types of sedimentary mineral deposits that are little understood, and as a result, many of them may have been overlooked. Examples of sedimentary mineral deposits already recognized are as follows:

- Redbed and black shale copper deposits, such as those in Zambia, White Pine in Michigan, and the Kupferschiefer of north central Europe
- Playa deposits, such as the MacDermitt, Nevada, mercury mine
- Kuroko-type deposits from deepwater environments.

An organized program measuring the contents of metals in sedimentary rocks would build a data base for recognizing and understanding these deposits.

In addition to their inherent scientific and economic interest, these sedimentary accumulations may represent a primary segregation process upon which secondary hydrothermal processes act to produce local mineable concentrations.

There are hundreds of thousands of feet of existing cores in storage in the United States that may be used to establish an initial data base. One or more portable laboratories that could visit core archives are needed. A rapid-scan instrumental, analytical technique is needed to examine intact cores without special preparation, such as the X-ray fluorescence logger mentioned above.

4. Program Costs

The only specific drill holes requested for the chemistry of the crust portion of the current program are those for the southeastern Arizona province. Most of the studies propose to use existing cores and to study core produced in the other parts of the continental drilling program.

The cost of cored NX (3.5 in., or 88.9 mm) drill holes is estimated to be $4,140,000. We propose that four core scanners be built, at a cost of $50,000 each, totalling $200,000. The three individual projects—grade-tonnage relations, metallogenic provinces, and sedimentary rocks—are estimated to cost approximately $1,500,000 each. The total program cost is therefore $4,700,000 for science and $4,140,000 for drilling.

C. Studies of Present-Day Mineral-Deposit-Forming Systems

The nature of active hydrothermal or geothermal systems and the manner in which they are related to ore-forming processes have long been one of the most intriguing questions in the science of mineral deposits. Often, the surface waters have very low metal content and, except for occasional mercury or other trace metals, show little or no evidence of significant mineralization in the recent past. Shallow drilling (to depths of a few hundred meters) in hot springs and geyser systems has shown little or no mineralization.

The deeper parts of active geothermal systems, instead, may be of great significance from the standpoint of ore-forming processes. Upper portions of these systems may reflect shallow convection cells of meteoric water, but in many instances their compositions suggest the existence of more concentrated fluids and of mineralization processes operating at depth.

Investigators of geothermal systems have been concerned with several questions bearing on energy production, such as the nature of the heat source and of the heat exchange between the source and the surface. Many of the same energy questions relate to problems of mineralization. In addition, the detailed chemistry of the fluids and the specifics of fluid-rock interaction, especially in the deeper, higher-temperature parts of these systems, assume great importance in mineralization processes.

1. Active Magmatic-Geothermal Mineral-Deposit-Forming Environments

a. Scientific objectives and problems to be solved. There is strong evidence that many existing mineral deposits have been formed in geothermal systems. The principal scientific objective of an investigation of currently active systems is a better understanding of the processes of deposit formation. Problems to be solved include identification of the critical processes, the interrelations of these processes, and the nature of the basic physical and chemical factors that control them.

b. Information needed. Much is already known about the near-surface geometry and geochemistry of geothermal systems. Deeper drilling is now required to gain an understanding of the region from about 1.5-km depth down through the "roots" of these systems. The following specific information is needed:

- The present temperature distribution, with depth, in various parts of geothermal systems
- Permeabilities, including fracture densities and orientations
- Grain size and other textural information
- Electrical resistivity
- Wall-rock mineralogy and major and minor element chemistry as a function of depth, temperature, and fluid composition
- Physical and chemical nature of the fluid as a function of depth, temperature, and mineral assemblage, including *in situ* fluid densities, pH, and fugacities of volatiles
- Rates and directions of fluid flow
- Oxygen, hydrogen, sulfur, and carbon isotope studies of all minerals and fluids

c. Holes required. It may be assumed that enough holes less than 1.5 km in depth already exist. We need additional holes that extend, if possible, far enough below the bottom of the hydrothermal circulation system to permit measurement of conductive heat flow at that level and the extent of influx of fluids from deeper levels, such as from underlying igneous bodies.

There are probably no existing holes that could profitably benefit from deep-drilling add-on experiments. First priority should be assigned to holes centrally located over the heat source; additional holes, such as those on the periphery, should be planned on the basis of data from the first hole. Two central holes might prove expedient for measurement of physical and chemical properties and flow direction.

d. Salton Sea geothermal field. The Salton Sea geothermal system is entirely within Pliocene and Quaternary sediments of the Colorado River delta at the north end of the Gulf of California. At the time of deposition these sediments consisted of sands, silts, and clays of uniform original mineralogical composition, but under the elevated temperatures and pressures of the geothermal system they are being transformed to low-grade metamorphic rocks of the greenschist facies. Temperatures within the explored geothermal system range up to 360°C at 7100 ft. ($\approx$ 2200 m). The wells produce a brine containing over 250,000 ppm (mg/l) of dissolved solids, primarily compounds of chlorine, sodium, calcium, potassium, and iron, plus a host of minor constituents.

Concentrated brine tapped by a deep well drilled for geothermal power near the Salton Sea in California deposited metal-rich siliceous scale at the rate of 2–3 tons/month. This iron-rich opaline scale contains an average of 20 percent copper and up to 6 percent silver present in bornite, digenite, chalcopyrite, chalcocite, stromeyerite, and native silver. The heavy metals in solution greatly exceed the sulfur content, on a modal basis. They are apparently derived from the sediments of the brine reservoir, being released from the silicate minerals in which they occur in trace amounts as metamorphism of the sediment proceeds.

Of the currently active geothermal systems, the one at Salton Sea has the highest temperature and the most metal-rich brines. According to White *et al.* (1963), it offers the "... fascinating possibility that this brine is man's first sample of an 'active' ore solution of the type that probably formed many of the world's economic concentrations of metals in the geologic past." Because of these unique characteristics, we consider deep drilling of the Salton Sea site to be our first priority.

We need to answer the following questions: In what form are the heavy metals carried in solution, and how do their solubilities change with temperature? Under what circumstances might the brine deposit its heavy metals in sulfide-ore deposits, and are there known ore deposits that originated from brines such as these?

An organized program of deep drilling is the only way to obtain the needed information on this unique ore-forming system. The highest-priority single borehole should be located in section 23, T13E, R11S, near the two wells previously drilled by O'Neil Geothermal, Inc. (Muffler and White, 1969). This hole should extend into the marginal parts of the underlying rhyolitic intrusive that lies at an estimated depth of 15,000 ft (5,000 m). Because of the very high temperatures and salinities expected at target depths, the scheduling of such a project will be dependent, to a large extent, upon technological capabilities. Because of the uniqueness of the information obtainable only from this system, however, the project should be undertaken as soon as technologically feasible.

e. The Geysers. In addition to its importance as a geothermal resource, The Geysers vapor-dominated reservoir may contain a natural laboratory for studying metal transport in hydrothermal systems. Three questions that relate to metal transport are potentially answerable at The Geysers:

- Is mercury being transported and/or deposited by the vapor-rich part of the system? In addition to its importance in connection with a class of mercury deposits, understanding mercury transport could possibly aid in avoiding inadvertent mercury production, which is a major environmental concern in geothermal energy production.
- Is the vapor-dominated system located atop a saline brine pool, or atop a stratified stack of liquids whose salinity increases with depth? If so, the brines may be metal-rich. Can one identify country rocks that have been the source of the metals, and are metals currently being deposited?
- Is boiling, as at The Geysers today, a temporary phenomenon in such deposits as the porphyry molybdenum and copper deposits? What are the transport mechanisms involved during even brief episodes of boiling?

The following information is needed from The Geysers for understanding mineral deposits: rock and fluid samples from the underlying water or brine pool; rock and fluid samples (some of which already exist) from the vapor-dominated region; profiles of temperature and fluid pressure with depth; *in situ* measurements of permeability, fracture density, and orientation; and, if the instrumentation is available in time, downhole measurements of pH, sulfide activity, and Eh (oxidation-reduction potential).

Needed boreholes include (1) one aimed at the most probable location of a deep brine pool (possibly with provision for drilling more than one exploratory hole to locate the target before drilling the instrumentation hole) and (2) one or more boreholes on the flanks of the geothermal field to evaluate sources of water and the chemical and isotopic nature of the feed into the system.

The major technological barriers are as follows:

- Management of the drilling fluid density as the drill penetrates deeper into a reservoir filled with low-density vapor

- Improvement of sampling techniques for high-temperature brines (the main reason for sampling the brines is because of their effectiveness in dissolving metals)
- Inability to chill a zone adjacent to the bit using the drilling fluid, because the heat transfer in the geothermal system is so efficient

Because the geothermal questions and the mineral deposit questions come together at The Geysers, there are multiple probabilities that an exploratory hole or holes deep at The Geysers would resolve many aspects of both sets of questions.

f. Yellowstone caldera, Wyoming. Many important base metal and precious metal ore deposits throughout the world are formed in direct association with large, predominantly silicic volcanic systems; these may or may not be associated with calderas. Examples are many porphyry copper deposits, most epithermal gold-silver deposits, and many base metal deposits such as those found in the San Juan Mountains, Colorado. Studies of this type of geological environment are obviously important in formulating theories of ore deposition.

The outstanding modern analog of such large silicic volcanic systems is the Yellowstone volcanic field in Wyoming. The chances of drilling into extremely hot, near-magmatic aqueous fluids is as good here as in any known geothermal system. Thus the possibility of penetrating the roots and studying a natural hydrothermal system that is closely associated with active mineral deposition is also as great here as in any area in the world. Because the fluids are probably less saline than those in the Salton Sea field and The Geysers, the technological problems of drilling into rocks at $500°C$ will be less at Yellowstone than at the other two sites. Very high temperature fluids will likely be intersected by drilling at relatively moderate depths at Yellowstone. Thus this hydrothermal area affords us the possibility of direct examination of present-day mineral deposition in a locality that is surrounded by many other economically interesting mines and prospects. Because of the northerly latitude, the distinction between meteoric and magmatic water can be very clearly delineated here by means of $^{18}O/^{16}O$ and deuterium/hydrogen analyses (in this respect Yellowstone is far better than the Salton Sea system). A wide range of salinities, temperatures, and pressures—as well as other characteristics, such as pH, total sulfur content, and total carbon dioxide content—is likely to be found in the hydrothermal fluids at Yellowstone.

Because of environmental concerns connected with drilling in a national park, there are probably going to be limited opportunities for drilling site selection at Yellowstone. Nevertheless, the hydrothermal area is so large that environmental and public concern can probably be met without significantly interfering with a program of deep scientific drilling. Of course, any such drilling at Yellowstone will have to be carefully monitored, and we must follow clear-cut safeguards to avoid any degradation of the geyser and hot spring systems. Any such drilling must also be totally dissociated from any conceivable implication of economic exploitation. It is absolutely essential that everyone connected with such a project from its inception make it clear that Yellowstone must not be exploited or disturbed in any permanent fashion. However, it is also important to recognize that because of such restrictions, we shall never be able to obtain information about this remarkably large and important hydrothermal system except through a major program of scientific drilling. A national scientific drilling program should therefore emphasize what a unique natural laboratory Yellowstone can provide. Such a concerted national scientific effort may, in fact, be one of the only ways to justify any type of deep drilling at Yellowstone. Because Yellowstone can never be drilled for commercial purposes, it is likely that only a national drilling program will enable scientists to obtain and study the wealth of geochemical and geological data in this natural laboratory.

Geological, geochemical, and geophysical data indicate that the Yellowstone Plateau volcanic field overlies a group of plutonic bodies of batholithic size, emplaced over a period of 2 m.y. The older parts of the pluton system are probably solidified but still hot. The youngest of these very large magma bodies has been emplaced during the last approximately 1.2 m.y. Two higher-level parts of this magma chamber formed a pair of adjoining ring-fracture zones, through which a climactic pyroclastic eruption of more than 1000 km^3 of magma occurred 600,000 years ago. The roof of the magma chamber collapsed along the two ring-fracture zones to form a 40 by 70 km caldera. Postcaldera rhyolites of intracaldera doming and three major sequences of rhyolitic lava eruptions have also occurred during the past 150,000 years, the latest being about 70,000 years ago. Geophysical evidence strongly supports the existence of molten magma in the present volcanic system at depths near 6-7 km. This implies the probable existence of very hot hydrothermal fluids at shallower depths. Temperatures as high as 237°C were measured in shallow research holes drilled in 1967-1978. Isotopic geothermometers and mixing models indicate subsurface temperatures commonly as high as 360°C, with salinities normally increasing downward because of decreasing local recharge and dilution. At deeper levels it is possible that more saline waters exist, perhaps gravity stratified, but higher in temperature.

Deep drilling in the Yellowstone caldera probably would first encounter 1-2 km of interlayered rhyolitic lavas and sediments, perhaps underlain by about 2 km of welded tuffs. At least in places, as much as 3 km of marine sedimentary rocks and Eocene andesitic sediments and lavas are underlain by a basement of Precambrian metamorphic rocks. However, intrusion of magma has cut out much of this sequence within the caldera. Judging by analogy with studies of fossil caldera systems, the deepest circulating meteoric waters probably will be found at depth along the caldera ring fractures. Magma

TABLE C-1   Estimated Costs of the Proposed Salton Sea, The Geysers, and Yellowstone Projects

Project	Number of Holes	Depth, ft (m)	Costs, $ million		
			Drilling	Research and Instru-mentation	Total
Salton Sea	1	15,000 ($\approx$ 5,000)	2.7	1.5	4.2
The Geysers	2	12,000 ($\approx$ 4,000)	2.4	1.6	4.0
Yellowstone	1	15,000 ($\approx$ 5,000)	2.7	1.5	4.2
TOTAL			7.8	4.6	12.4

bodies emplaced at shallow depths within the ring-fracture system may therefore be likely drilling targets, assuming these can be identified from geological and geophysical information. Depending on specific locations, the depth to the solidified top of the magma chamber may be 3–7 km and depth to still-molten magma perhaps 6–7 km.

The sensitive nature of drilling in an area considered a national treasure must always be in the forefront in considering such a project. Drilling should not begin until the technology of completing holes under such adverse high-temperature hydrothermal conditions is reliably established. Time on this drill site should not be wasted in developing high-temperature drilling technology. The Yellowstone program would therefore be among the later parts of a national drilling program. This site should, of course, be selected to offer the minimum environmental insult and disturbance of park activities that is compatible with scientific objectives.

The beauty and uniqueness of Yellowstone Park result from the igneous activity that underlies the park. The detailed knowledge of the subsurface environment would complement existing park information and make Yellowstone a more meaningful national park.

g. Program costs. Because of the unusually high temperatures and corrosive environments that will be encountered in drilling to the deeper parts of active geothermal systems and because much of the required technology remains to be developed, present estimates of drilling costs at best must be regarded as crude. Similarly, the cost estimates of scientific instrumentation for downhole measurements are also crude. Thus, with a great deal of reservation, costs of the proposed Salton Sea, The Geysers, and Yellowstone projects are estimated in Table C-1.

2. Metal-Rich Brines in Deep Sedimentary Basins

The surprising discovery by Carpenter et al. (1974) of metal-rich formation brines in the Mississippi Embayment with low-metal, H_2S-bearing waters in underlying formations suggests that the deposition of Mississippi Valley-type lead-zinc deposits could be an effect of late diagenesis. Further study of these brines and the search for additional examples promise to clarify the origin of this class of previously cryptic ore deposits.

The major scientific objectives are (1) to characterize the brine compositions, (2) to develop an understanding of the origin of both the metal-rich and the H_2S-rich brines, and (3) to test whether any isotopic and chemical signatures of the brines are evident in mineral deposits of the Mississippi type. Given the composition of the brines, can one predict the reactions upon mixing, with or without the buffering of adjacent carbonate rocks?

The information needed is the composition of the major elements, selected trace elements, and isotopes from carefully selected samples of Mississippi Embayment brines and of the shales and other rocks that might have been leached of metals. Present temperatures and the temperature history of the rocks (from kerogen and pollen maturation) are also needed. An example is the southeast Missouri lead belt, where lead may possibly have been leached from the overlying Davis shale. Thus the hydrological flow regime becomes an important factor in assessing transport and possible mixing behavior of the brines. Careful records of pressure from drill-stem tests will be needed.

A second-category question is: Is the Mississippi Embayment occurrence unique? A broad sampling of the deep holes now being drilled for natural gas could be based, first, on water from existing drill-stem tests and, second, on water sampling done for purely scientific purposes where high temperatures and low electric-log resistivities indicate high-salinity brines. This will be a unique opportunity during the next decade, because the deeper and hotter portions of sedimentary basins have long been avoided during the search for oil; now, with the progressive deregulation of natural gas prices, these deep, hot areas are going to be systematically drilled. A decade ago there were few wells, but a decade from now it will be too late.

The holes needed can be supplied entirely from existing drilling for oil and gas. At most, water sampling from selected formations would have to be funded by the scientific budgets.

The first priority would be additional samples, collected under careful controls, from the Mississippi Embayment for more detailed chemical and isotopic study. The second priority would be water samples from wells drilled for deep gas exploration, such as in the Anadarko Basin in Oklahoma. A third priority would be the central areas of the many basins known to have high salinities. Another approach is to resample known aquifers of high-salinity brines and examine in detail the trace metal content in relation to shale deposits. The mechanism of ore deposition has to be examined in three stages: mobilization, transportation, and deposition. All of these processes must be related in one way or another in high-brine systems.

    a.   Program costs. No specific drill holes are proposed for studying metal-rich brines in sedimentary basins. Vast numbers of oil and gas wells are being drilled in the region of interest that could be used in the investigation. Still lacking is an adequate technology for downhole sampling of pore waters, but once this is developed, as we believe it can be, the costs will be those of sampling and of scientific studies. We estimate that the project could be completed by four people over a five-year working period at a cost of $1 million.

3. Role of Pores and Microfractures in Material Transfer

A related, but separate, problem of great significance can be studied only in present-day hydrothermal- and sedimentary-basin pore-fluid systems. This is the question of how chemical elements and complexes, both organic and inorganic, move through volumes of rock. The problem is fundamental to an understanding of both mineral concentration processes and petroleum migration. A critical portion of this problem may be addressed by a study of material transfer in cores through pores and microfractures. Equally important is a study of the various causes of microfracturing, particularly the role of internally generated fluid pressure.

The research proposed differs from geothermal inquiries in that it focuses on the *initial mobilization* of chemical elements and complexes, as well as on their subsequent transport and precipitation. Specifically, do interconnected rock pores and fractures exist over a range of sizes, so that essentially all of the rock matrix is in chemical communication with the saturating fluid? It will not be necessary to drill holes specifically for the purposes of this research. Rather, core and fluid samples from existing and planned holes may be studied. Samples of particular interest include those from active hydrothermal areas, pervasively mineralized areas, and overpressured shales.

The objective of these studies will be to determine the size of the pores and fractures present and their spacing, orientation, and interconnection and to determine the relation between the composition of the rock matrix and that of the saturating fluid. The latter studies should concentrate on the chemical reactivity of the rock surface, on the differences between migratory and nonmigratory fluids, and on the composition of fluid inclusions and the solids that fill annealed fractures. Causes of microfracturing can best be investigated through the acquisition of *in situ* data on pore pressures and on pore-fluid composition and through artificial fracturing experiments in the hole.

Finally, it will be necessary to carry out surface and borehole geophysical studies to relate the properties of the sampled material to those of rocks in the broad vicinity of the hole. This may be accomplished both by standard logging of the hole at the time of drilling and by postdrilling borehole experiments designed to delineate the three-dimensional density, electrical conductivity, and seismic velocity structure near the hole. In particular, geophysical experiments with active seismic and electromagnetic induction sources at various positions on the surface, and receivers down the hole, should be carried out to detect possible anistropy or azimuthal variation in electrical conductivity or seismic velocity caused by preferred orientation of the fractures. Experiments for this purpose have already been carried out in shallow (3000 ft, or ≈ 1000 m) holes, and the methods for the interpretation of the data are well established. However, advanced downhole instrumentation may be necessary to probe the deep sections of active hydrothermal systems.

a.   Program costs. Because separate drill holes are not proposed for this project, all objectives may be reached by using holes drilled for other purposes. We estimate that experiments will have to be performed in holes as deep as 10,000 ft (≈ 3,000 m) and that some equipment and technology development will be needed. Each hole will require 4 man-years of study, and the cost, including equipment, is estimated at $500,000 per hole. We believe that at least five holes will have to be studied, costing about $2,500,000.

D. Studies of Mineralized Systems

This section is aimed not at prospecting, but at a more thorough understanding of the processes of mineral formation. Questions bearing on prospecting are inevitably involved (e.g., the nature of the tops, bottoms, and middle portions of typical ore deposits), but the motivation for study is the elucidation of the nature of the physical, chemical, petrological, and tectonic environments involved in mineral formation.

We suggest that initial targets be deposits formed by magmatic-hydrothermal systems. In part, this selection is made because the geometry of the deposits and of the altered rock volumes in which they are enclosed seems to be less complicated than the geometry of deposits such as stratiform copper, lead, and zinc deposits and also because the magmatic-hydrothermal deposits are clearly the products of fossil magmatic-geothermal systems. This selection does not downgrade the extreme importance of other classes of mineral deposits. Rather, it recognizes the higher probability of success for an initial drilling program, if it is focused on magmatic-hydrothermal deposits.

Drilling in mineral districts in the past has been typically shallow (less than 5,000 ft, or ≈ 1,500 m) because mining companies are unable to justify drilling much below the zones of economic mineralization. As a result, our understanding of the three-dimensional crustal structure and chemistry, as well as the structures of the mineral deposits themselves, is, in each instance, poor. Very little drilling to the depths typically attained in petroleum exploration has been done. We therefore stress the need for large numbers of intermediate-depth holes (5,000–10,000 ft, or ≈ 1,500–3,000 m) in several types of mineral deposits or mineral districts, before or at least concurrent with any extremely deep drilling.

There are many scientific objectives for drilling selected mineral deposits, but four major questions stand out:

- Are deposits single entities, or are they components of larger systems of deposits? Do the so-called epithermal gold-silver deposits grade at depth into one or another of the classes of basement deposits? Are some classes of massive sulfide deposits genetically related to porphyry-copper deposits? A great deal of evidence, such as mineral zoning, exists to support the concept that most deposits represent only parts of systems, which may have vertical dimensions of 15,000–30,000 ft ($\approx$ 5,000–10,000 m). No entire system has ever been explored, because attention has always been focused on the actual ore bodies.
- Do significant zones of leaching occur in the rocks that underlie mineral deposits? The source or sources of constituents in magmatic-hydrothermal deposits remain, in most cases, obscure. Presumably, they come from the deeper parts of the hydrothermal circulation system, but because the deeper zones have never been studied, we cannot be certain. Deep drilling is the only way to get the fresh rock samples needed to answer the question.
- The origin of the water in magmatic-hydrothermal systems is always difficult to establish. Isotopic evidence suggests that both magmatic and meteoric sources are important at various stages of deposition, but the answers to questions about the degree and extent of vertical mixing of such waters and their relative importance in different portions of the hydrothermal systems are incomplete. Deep drilling can provide the samples to answer these questions.
- What do the deepest portions of a deposit-forming system look like? Do they extend deep into the parent magma body, or do they end in highly leached zones in the country rock surrounding the magma chamber?

We recommend that four different types of mineral deposits be drilled. In each, the deposits have been well-studied geologically, but specific drill hole locations will be selected following complete evaluation of the surface and mine-mapping information already available. Each hole will have similar objectives, and in each case, the same kinds of studies are to be performed. These studies are as follows:

- Petrology
- Kinds of mineralogical alteration
- Structural relations
- Absolute ages of all rock units
- Geochemical and isotopic trends
- Chemistry of present pore fluids
- Fluid-inclusion studies as a guide to past ore fluids

Additional measurements and studies may be performed as appropriate for specific holes.

1. Tonopah, Nevada

The Tonopah, Nevada, gold-silver deposit is an excellent, well-studied example of a typical epithermal deposit in a volcanic terrane. The mine is now closed, but approximately $150 million in gold and silver was mined here before World War II. The geology of the district was described by Nolan (1935). The ore occurs in steeply dipping quartz veins within a highly faulted and fractured dacite-rhyolite volcanic pile that has suffered intense hydrothermal alteration. The ore-bearing veins are confined to a subhorizontal zone about 500 ft ($\approx$ 150 m) thick that forms a blanket draping over the central part of the district. The apex of this blanket defines the center of the district, and it provides a clear-cut drilling target. Above and below this blanket, the veins are not economic, so drilling and mining have penetrated to depths of only about 1500 ft ($\approx$ 500 m).

The deposits are known to have been formed by meteoric-hydrothermal waters, as a result of meteoric groundwaters forming a hydrothermal convective system above a hidden intrusive stock. The salinity of the depositing waters (1 percent sodium chloride equivalent) and the temperatures (250°- 300°C) were low. An intrusive igneous body is necessary to drive such a hydrothermal system, and Nolan postulated that such a stock underlies the center of the district; it is likely that a centrally located drill hole would intersect the stock at a depth of about 1-3 km. Once the stock is reached, the type of mineralization encountered will guide deeper drilling, but the stock certainly should be drilled to a sufficient depth to determine whether it is the source of the gold and silver in the Tonopah veins. One or two flanking drill holes will be required to establish the overall three-dimensional mineralization pattern.

The principal scientific objectives are as follows:

- To establish the presence and nature of the intrusive body at depth
- To determine whether any magmatic water was present in the ore-forming fluid (perhaps increasing in amount with depth)
- To establish the mineral and alteration zoning pattern beneath a typical epithermal deposit
- To trace the vein systems within the stock
- To determine whether the stock was the source of the metals or whether they were leached from the adjacent country rocks by the hydrothermal systems
- To determine whether any gold-silver, base metal, or porphyry-copper mineralization underlies the explored zone of gold-silver mineralization

2. Tintic, Utah

A historically important and common type of magmatic-hydrothermal mineralization in the western United States and the American cordillera is the silver-lead-zinc association. One of the major questions in the geology of mineral deposits has always been where these deposits lie in crustal position and in pressure-temperature-composition environment in relation to the epithermal deposits and the porphyry-copper deposits. The Tintic district consists of rich veins and mantos (flat-lying bedded deposits) in carbonate host rocks with several features characteristic of porphyry-copper districts (i.e., a volcanic setting, locally strong sericitic and argillic alteration, and the occurrence of breccia pipes and pebble dikes). At least four carefully positioned drill holes are required for adequate deep investigation and a highly desirable better understanding of this district. This is especially important because of the porphyry-copper mineralization seen in the Salt Lake district, which lies immediately north of Tintic and shares many geological features in common with it. The geology has been well-studied by Morris and Lovering (1961) and a number of earlier investigators.

3. Butte, Montana

The Butte, Montana, base metal vein system emplaced into the Boulder batholith is one of the most extensively studied mineral deposits in the world. The wealth of information obtained during the past 70 years of study on this deposit has provided perhaps the best three-dimensional picture yet developed of a hydrothermal vein deposit (Meyer *et al.*, 1968). The three-dimensional zoning of both the vein systems and their alteration halos is relatively simple, and the deepest has been mapped in detail in the extensive mine workings over a vertical distance of 1.5 km. Convectively circulating meteoric-hydrothermal fluids have been shown to have dominated during the latest stages of ore deposition, but the question of whether magmatic water was important at greater depths and/or early in the deposit-forming process remains unanswered. Thus this deposit also presents an especially favorable opportunity for studying the deepest levels or "roots" of a gigantic meteoric-hydrothermal system. The metal zoning at Butte (molybdenum in the deepest levels, copper in the central zone, grading outward into a zinc-manganese-silver peripheral zone) is one of the world's classic examples of such district zoning; characterization of the deeper zonation pattern would complete this picture and would be significant in understanding the ore genesis.

As at Tonopah, one of the principal scientific objectives at Butte is to determine whether the copper and other metals have been leached from a source rock at depth (perhaps from a deep porphyry-copper deposit) and then reprecipitated in the overlying vein system or whether they have been directly supplied from a magma at depth. Because of the many man-years of scientific effort already undertaken at Butte, only a modest amount of deeper drilling would significantly augment the most complete three-dimensional picture of any known ore deposit.

Other specific questions to be addressed by the deeper drilling at Butte include: What is the size, shape, and chemical composition of the intrusive stock that drove the meteoric-hydrothermal circulation? Is this stock the source of the ore metals, or did it simply provide the heat necessary to drive a gigantic leaching system? How do the veins change in character with depth, and what does the bottom of such a gigantic vein system look like? Since the meteoric-hydrothermal convective system has already been shown to extend over a vertical distance of at least 2 km, might it be possible to determine whether it actually extends to depths of 5-7 km? Detailed characterization of a geothermal system of such size within a granitic batholith would be of great scientific importance, even if there were no accompanying mineralization. If a magmatic system existed in the earlier stages of mineralization, how and over what dimensions was such a system modified by the superposition of the later meteoric-hydrothermal system?

It is probably not going to be necessary to drill completely new holes for the scientific purposes outlined above. Arco-Anaconda has already developed a program of intermediate drilling to depths of 8,000 ft ($\approx$ 2,600 m) or more, and it should be possible to use or extend some of these holes. However, at least three deep holes will be required, and these should extend to depths of at least 10,000 ft ($\approx$ 3,000 m). At Butte the overall scientific return will be markedly enhanced if drilling is extended far enough to penetrate the entire zone of hydrothermal circulation, even if such effects should extend to depths of more than 15,000 ft ($\approx$ 5,000 m). Such a study would provide the clearest and most complete knowledge yet attainable on the permeability and hydrothermal circulation deep within a granitic batholith.

4. Santa Rita, New Mexico

One of the more completely studied porphyry-copper deposits of the southwest United States is the Santa Rita ore body. It consists of a composite stock containing both equigranular and porphyritic quartzmonzonite components intruded into a dominantly Paleozoic carbonate section. Chalcopyrite, both disseminated and controlled by fine fracturing, is the major ore mineral. Primary mineralization is of highest grade in areas of potassic and mixed potassic-sericitic alteration and extends into skarn ore bodies peripheral to the complex. In addition, lead-zinc mineralization occurs at greater distances in silicated limestones, controlled primarily by grandioritic dike contacts. The alteration-mineralization pattern at Santa Rita follows, in general, the classic porphyry-copper model much discussed in the recent literature. At Santa Rita, significant mineralization apparently extends to considerable depths, but trends in petrology, structure, fracturing, alteration type, and the possible appearance of new phases (e.g., anhydrite) are not known. In short, the nature of this transition to the presumed subjacent batholith, at Santa Rita and in all of the porphyries, is yet to be established. The presence of Santa Rita within the unique copper-metallogenic province of southeast Arizona and southwest New Mexico and the enigma of the existence of this province are an additional reason for its selection.

Although all of the porphyry-copper deposits have certain features in common, they also have individual characteristics that represent variations on a theme. The reasons for these variations are obscure and must involve factors such as original depth of intrusion, original composition, the numbers and timing of successive porphyry surges, and the nature and strength of the superimposed convective, meteoric environment. Drilling several deposits to better elucidate these points is not yet justified, but by careful selection and drilling of at least one of the better-understood bodies, such as Santa Rita, we can at least maximize our return. Once this project is complete, the way would be open to plan an investigation of an additional deposit; at present a most logical one would be Bingham, Utah.

5. Program Costs

Each deposit selected should be studied in two phases. The first phase should involve up to four wire-cored major NX (3.5 in., or 88.9 mm) or BX (2-7/8 in., or 73 mm) holes, drilled to a depth of between 5,000 and 10,000 ft ($\approx$ 1,500–3,000 m). If holes have to be drilled, costs for each deposit will range from \$345,000 to \$900,000. In many instances, however, we anticipate that shallow holes already drilled by mining companies can be deepened to reach the target zones, thereby reducing the total cost. Thus a realistic cost for each deposit would average about \$600,000. Phase 2 of each drilling project would involve at least one deep hole in the 15,000- to 20,000-ft ($\approx$ 5,000–7,000 m) range for a cost of about \$3 million per site. Thus if four sites are followed through phases 1 and 2, the expected drilling costs will be as follows:

Phase 1	\$ 2,400,000
Phase 2	\$12,000,000
Total	\$14,400,000

The scientific studies attending the drilling and examination of drill cores are difficult to estimate for this program. The concept of the program is so revolutionary that a great many unexpected discoveries may be anticipated. The best we can do is estimate the cost of the initial studies of the cores and recognize that many later studies will follow. We estimate that at least \$500,000 per hole will be needed to study and examine the cores. The science program is estimated at \$10 million for both phase 1 and phase 2.

Appendix D
Earthquakes

D EARTHQUAKES

I. Earthquake Problems and Current National Needs

Because of the awesome potential of earthquakes for causing loss of life and extensive property damage, many persuasive arguments can be made for studying seismic processes. But in terms of truly urgent national needs, three subject areas stand out in which the lack of fundamental knowledge concerning earthquakes has seriously impeded programs of generally accepted high national priority. These are: (1) the finding of suitable sites and techniques for isolating nuclear waste products, (2) the siting and earthquake-resistant design of nuclear-generating facilities, and (3) the siting and design of major dams. Particularly (but not exclusively) in the western United States, almost every site selected for such facilities has become embroiled in controversies concerning seismic hazards. These controversies, most of which represent honest differences of scientific opinion, in large part reflect our basic ignorance of seismic phenomena. Only through a better basic understanding of earthquakes—why, how, and where they occur—will we be able to select sites for and design such critical facilities economically and with the confidence in safety that the public rightfully expects.

II. Scientific Background and Applications

A. Background

1. Distribution of Seismicity

Before the concept of plate tectonics, no satisfactory explanation was available for the world-wide distribution of seismicity. Plate-tectonics proponents hypothesize that the surface of the earth is broken into a series of plates that are in motion relative to each other. In terms of plate tectonics, a large fraction of the earth's seismicity may be associated with interactions between plates at their margins. The extensive seismic activity on the San Andreas fault in California, for example, is certainly the result of lateral sliding between the Pacific and North American plates. Resistance to this sliding results in great earthquakes. The great earthquakes in Alaska are directly associated with the descent of the Pacific plate into the mantle along the Aleutian trench. This process is the basic cause of a major fraction of the largest earthquakes recorded.

Although plate tectonics provides an understanding of the seismicity in general terms, many detailed aspects remain unexplained. Earthquakes in California are not restricted to the San Andreas fault. Displacements occur on many interconnecting faults. In fact, significant seismicity extends throughout the western United States. The relative probability of major earthquakes on these secondary faults can only be approximated. In general, this seismicity may be explained in terms of a broad region of interaction between the Pacific, North American, and Juan de Fuca plates. On the basis of present knowledge, great earthquakes could occur virtually anywhere in the region.

In comparison with the western United States, the eastern United States is relatively free of seismic activity. Nonetheless, major earthquakes did occur near New Madrid, Missouri, in 1811 and 1812, and near Charleston, South Carolina, in 1886. Other areas of minor seismic activity are scattered throughout the eastern region. The causes of this seismicity are inadequately understood; therefore the probabilities of earthquakes in particular areas can be based only on the historical record.

In terms of plate tectonics the Hawaiian Islands should not exist. These Islands are the result of extensive volcanism near the center of the Pacific plate. Associated with the volcanism is a significant level of seismicity. Although several hypotheses have been proposed to explain the Hawaiian Islands, none has gained wide acceptance.

2. The Earthquake Mechanism

Earthquakes are generally associated with the release of accumulated stress by catastrophic rock failure. Although this is undoubtedly one mechanism for producing earthquakes, it is likely that there are others. Earthquakes occur at depths of 700 km. At depths greater than 20 km, it is very doubtful whether brittle (or catastrophic) rock failure can occur; satisfactory alternative explanations for deep earthquakes are therefore required. These alternative mechanisms may also be operative at shallower depths.

The San Andreas fault is probably the best place in the world to develop an understanding of the fundamental earthquake processes. The fault is well exposed at the surface and is largely the result of lateral sliding. Long before the plate-tectonics concept was developed, Reid (1910) explained the 1906 San Francisco earthquake in terms of the "stick-slip" mechanism. The two sides of the fault stick together until a critical stress is reached, at which time slip occurs and results in an earthquake. Laboratory and theoretical studies have reproduced this type of behavior. However, at least part of the San

Andreas fault appears to be freely sliding, without stick-slip. The reason that different parts of the fault behave in different ways is not understood. Likewise, the critical level of stress required to produce catastrophic slip is not understood. Nor are the conditions that determine the extent of an earthquake rupture understood. Answers to these important questions are fundamental to predictions of the expected magnitudes of earthquakes.

3. Earthquake Prediction

The strategy that has guided earthquake prediction research to this point is primarily synoptic or data-gathering, supplemented by some basic studies and topical investigations. This strategy has had some success in the People's Republic of China, where such empirical studies have led to successful predictions of earthquakes. However, for seismic zones in North America the strategy suffers because there are not enough moderate-to-large earthquakes in any given area to provide an adequate data base within a reasonable period. This is particularly true of seismic zones in the central and eastern United States.

As an alternative to instrumenting a large number of areas with the hope of "capturing" events for the testing of prediction techniques—which would require more money than is available for earthquake prediction research—a series of direct borehole measurements in active fault zones might determine which parameters are most significant to the development of a physical model for earthquakes that would, in turn, predict the character of premonitory signals preceding earthquakes. Empirical data gathered by the monitoring networks would still be needed to test these predictions, but testing a soundly based hypothesis would require only a relatively few events as compared with the number required to develop a prediction capability primarily on an empirical basis.

An empirical scheme for predicting earthquakes will have some successes, but it also seems certain to result in failures as well. Until the nature of the mechanical system that produces large earthquakes is well understood, earthquake prediction will be a frustrating exercise—one that could even self-destruct through society's intolerance of expensive errors.

B. Applications

1. Seismic Hazard Assessment

The occurrence of large earthquakes along the San Andreas fault system poses a terrible human and economic hazard to major population centers in California. A major effort under the National Earthquake Hazard Reduction Program is aimed at a better understanding of the tectonic processes and material properties of the crust in this system. Within this program, preliminary drilling has been conducted (or is planned) to measure material properties and conditions of temperature and pressure associated with portions of the fault that are thought to be locked together as well as some that are sliding. These measurements will comprise a data base, not only for ascertaining the mechanisms that drive the plate motion along the Pacific coast, but also for providing a means of assessing relative levels of hazard associated with state of stress in the crust for various parts of the fault system. The occurrence of moderate-to-large earthquakes in other parts of the country is also recognized as a significant hazard, the extent of which may be generally estimated from seismicity patterns of the past. In general, these studies have serious drawbacks, stemming in large part from the lack of data covering sufficiently long time periods. A regional study of *in situ* stress in the continental part of the North American plate could go far toward identifying those areas of greatest risk, which would therefore require careful monitoring. As a corollary, the study could also be used to define areas in which crustal stress levels represent a negligible seismic hazard and where currently stringent requirements for the construction of power plants and other critical facilities could be safely relaxed. Cost savings associated with reductions in requirements for earthquake-resistant construction could be tremendous.

2. Induced Seismicity

After a number of surprises in recent years, most geologists and geophysicists agree that some earthquakes have been induced or triggered by works of man. This has happened in several ways: by underground fluid injection, by large underground explosions, and by the filling of reservoirs. Reservoir-induced seismicity was first recognized in the 1930's with the filling of Lake Mead (Carder, 1945). In the 1960's, four major earthquakes occurred at reservoirs (Hsinfengkiang, China, 1962, m_b = 6.1; Kariba, Rhodesia-Zambia, 1963, m_b = 5.8; Kremasta, Greece, 1966, m_b = 6.3; Koyna, India, 1967, m_b = 6.5) (Simpson, 1976). Two of these events—Kremasta and Koyna—resulted in death, injuries, and extensive property damage, and in two—Hsinfengkiang and Koyna—the dams also sustained damage.

From the geological-seismological point of view, two major problems remain unsolved: What is the mechanism by which reservoir-induced earthquakes occur, and why have earthquakes been triggered by some reservoirs and not by others? In general, stresses caused by the reservoir are superimposed on a preexisting stress regime; whether or not earthquakes occur depends on the interaction among the complete tectonic, geological, and hydrological environments. In a region with sufficient preexisting stress, earthquakes may be related either to an increase in the vertical load of the mass of the reservoir or to a decrease in effective stress caused by increased pore-fluid pressure in fault zones that may be present in the vicinity of the reservoir (Raleigh et al., 1976; Hubbert and Rubey, 1959).

To date, work on this problem has concentrated mainly on data collected with local seismic networks or in geodetic surveys. Now it is of crucial importance to begin a program of direct monitoring of parameters that may be related to earthquake inducement. Such measurements should be initiated before a reservoir is constructed and should continue after it is filled. The problem of reservoir-induced earthquakes cannot be overestimated—millions, if not billions, of dollars are now being spent to design large dams so that they will be resistant to earthquakes, and a careful research program incorporating borehole measurements to determine the physics of earthquake inducement may show this expenditure to be unnecessary under some geological and hydrological conditions.

3. Seismic Verification

Quite aside from problems involving earthquake hazards, seismological studies have been of critical importance in recent years in connection with the differentiation between earthquakes and underground nuclear explosions. A better understanding of the continental crust in terms of the propagation and attenuation of seismic radiation in distinct geological and geophysical regions would allow much more confident monitoring of treaty compliance. Borehole-to-borehole and borehole-to-surface studies in the near and far field would contribute markedly to increasing our understanding in this area.

III. Role of Drilling in Understanding the Earthquake Process

During the past five years, research into the earthquake mechanisms and the possibility of prediction has made considerable progress. Most of the research employed so far has been based on geophysical observation at the surface of the earth. Although many promising leads have developed, it has become abundantly clear that the process is much more complex than our simple models indicate, and the subsurface physical properties are also very much more complex than those of our current models. Resolution of the details of these more complex models of the fault region requires that some observations be made in the third dimension, i.e., in depth.

It is by no means clear whether the difference in the behavior of the locked and the creeping segments of the San Andreas fault arises from differences in the state of stress in these segments and in the time rate of change of stress in the segments or in the physical properties of the rocks in the respective segments. Measurements of the stresses or the time rate of change of the stresses cannot be made at the surface or even at shallow depth, for the surface observations have shown that there are significant changes in physical properties with depth. Thus drilling is necessary for several reasons:

- It provides an opportunity to obtain samples from the neighborhood of the fault for the study of physical properties in the laboratory and comparison of these properties with those in places that are some distance from the fault.
- It provides an opportunity for *in situ* measurements of the stress, its variation with depth, and the time rate of change of stress in magnitude and direction.
- It provides an opportunity for measurements at depths remote from the effects of the free surface, weathered zone, or terrane.
- It provides an opportunity for making measurements from sources on the surface to sensors in the hole, and *vice versa*, or hole-to-hole studies. This provides much more effective sampling of the physical properties of the rocks in the neighborhood of the fault in three dimensions than is possible from surface observations above. Use of this information greatly facilitates the interpretation of the surface observation.
- There are measurements that can be made only in boreholes. Pore-fluid pressure is known to have an important role in determining the response of rock to stress, and it is possible that pore-fluid chemistry is equally important. The need for study of the chemistry of the pore fluids suggests that in many instances air drilling techniques should be used.

Understanding the mechanics of reservoir-induced earthquakes also cannot be achieved without observations at depths near and below the reservoir.

IV. Relevant Observations

To improve our understanding of the distribution and causes of earthquakes, extensive observations are required. Many observations are best made in boreholes; some can be made nowhere else:

- Stress. Earthquakes occur to relieve stress; therefore direct measurements of stress are expected to provide essential information on the conditions leading to major earthquakes. Stress measurements in near-surface rocks tend to be strongly influenced by the presence of local faults, fractures, and other near-surface inhomogeneities. Stress measurements in deep boreholes appear to give satisfactory regional stress patterns and levels. Extensive measurements of stress as a function of depth throughout the United States should contribute to a fundamental understanding of conditions leading to earthquakes. Stress measurements adjacent to the San Andreas fault may provide information on the risk or occurrence of a great earthquake, as well as information on the various mechanisms operative along the fault.

- Temperature and heat flow. Because faulting is known to generate heat, measurements of heat flow may provide important constraints on fault processes. At present, heat flow measurements adjacent to the San Andreas fault are believed to place an upper limit on the shear stress across the fault. Clearly, more extensive heat flow measurements adjacent to and within the San Andreas fault—as well as other faults—will provide important information.

- Fluid pressure and permeability. In many theories for the generation of earthquakes, water pressure plays an important role. High water pressure may reduce the effective stress and hence the frictional resistance along faults, which may allow earthquakes to occur at lower stress levels. Measurements of water pressure would test these hypotheses if carried out in areas of different seismic behavior. Such measurements can be carried out only in boreholes. *In situ* permeability measurements are extremely important to test the validity of such hypotheses as dilatancy-diffusion (preceding earthquakes) and fluid flow as the cause of time-dependent phenomena such as aftershock and foreshock sequences. Such measurements are also important in areas of potential seismicity induced by reservoir impoundment.

- Seismic observations. Seismic observations provide a wide range of information about the occurrence of earthquakes. Small earthquakes sometimes delineate faults. Moderate-size earthquakes may be evidence for a high regional state of stress, indicating a high seismic risk. Although most seismic observations may be made using surface seismographs, instruments in boreholes provide a number of advantages: They are remote from the high-surface-noise level. They are particularly suited for obtaining high-frequency signals and determining the location of earthquakes of very low magnitude.

- Velocity and attenuation measurements. Seismic velocity determination has been the fundamental tool for extending structural knowledge of the earth to depth. Such measurements in boreholes are particularly important for calibrating field surveys and for extending information gained in boreholes over greater areas.

- Strain measurements. Surface strain occurs while the stress is building up before an earthquake. Extensive measurements of surface strain are being conducted. As with the surface stress measurements, surface strain measurements are adversely affected by the inhomogeneities of the surface rocks. Up to now, very few borehole strain measurements have been made.

Such measurements may have important advantages, at least in terms of the vertical component of strain.

- Material properties. Estimates of the strength of subsurface fault zones vary by at least an order of magnitude. Recovery and analysis of such material by coring will greatly benefit the understanding of fault zone processes.
- Pore-fluid chemistry. Recent theories of earthquake processes incorporate processes involving chemical changes, and such determinations are especially important. Determination of the origin of subsurface waters in active fault zones will also be of substantial benefit.

V. Present Effort and Programs

A major scientific effort is currently being carried out or supported by the U.S. Geological Survey in its capacity as the lead federal mission-oriented agency concerned with earthquake-hazard reduction. In addition, other agencies and groups throughout the country are currently conducting drilling programs that are devoted at least in part to earthquake studies, such as the drilling of numerous shallow holes to define the locations and geometries of fault planes.

VI. Need for Coordination, Instrument Development, and Manpower

Appreciable expansion of drilling programs for scientific purposes will require improvements in interagency coordination and information exchange with the scientific community, development of new instrumentation, and enlarging the pool of adequately trained people.

A. Coordination

If holes are to be drilled for multipurpose and/or interdisciplinary use, lead time is a critical factor. The minimum time required for planning experiments like hydraulic fracturing ranges from six months to a year or more, for most universities and federal and state agencies. Even longer lead times are desirable. The following information is needed to determine whether a hole might be appropriate for add-on experiments:

- Primary objective for drilling the original hole
- Location and target depth

- Diameter of hole (maximum and minimum if variable)
- Type of drilling fluid
- Sample, core, and logging plans
- Disposition planned (such as cased, uncased, capped, backfilled)
- Summary of lithology

B. Instrumentation

Many needs for downhole instrumentation are common to the several continental drilling programs, and these are discussed in detail in Appendix E. Here we emphasize particular developments that will be critically needed for adequate understanding of earthquake physics.

- Improvements in the precision and economy of determinations of *in situ* stress by the hydraulic fracturing technique.
- Invention and/or development of alternative and potentially superior techniques for downhole stress measurements.
- Encouragement of development of the high-pressure coring device that preserves the *in situ* state of the rock and its fluid content.
- Improvements in resolution and data acquisition of the borehole televiewer.
- Development of a downhole instrument for long-term monitoring of local shear strain (deformation of the borehole).
- Development of a downhole instrument for long-term monitoring of stress changes, preferably the whole tensor but at least the average stress normal to the borehole.
- Further development of satellites for long-term telemetry to reduce transmission costs.

C. Manpower

Petroleum, mineral, and mineral fuel exploration in the coming decade will require much more manpower—with relevant training in drilling and its applications—than is now available. The proposed scientific drilling program will provide extensive opportunities for the necessary training. To accomplish this objective, it will be necessary to involve the existing university research groups in proposed drilling projects and to encourage the formation of new research groups.

One effective way of developing complementary research in universities is to provide external grants. This should be a part of any major government drilling effort. We feel that a significant fraction of this external support should involve university research groups in the actual drilling activities. In these instances, support should be contingent on the actual participation of graduate students and/or faculty in the drilling activities.

VII. Augmentation of Ongoing Programs

We recommend that all major drilling dedicated to the earthquake-hazard program be coordinated with the U.S. Geological Survey. This recognizes its designated lead agency role among mission-oriented federal agencies as specified in the Earthquake Hazards Reduction Act of 1977, but we by no means wish to exclude the possibility of the instigation of their own programs by other groups, individuals, or agencies under special circumstances.

We strongly recommend that funding for existing programs that are directed at an understanding of the earthquake process be increased to establish a firm foundation for earthquake prediction and hazard reduction. Expanding the depth and number of holes, at a reasonable rate, in areas of critical interest will incur costs roughly as follows:

1980	+ $ 4 million
1981	+ $ 6 million
1982	+ $10 million
1983	+ $15 million
1984	+ $25 million

Funding in 1983 and 1984 provides for at least one slant hole to a depth of approximately 10 km, as previously recommended at the Ghost Ranch Workshop.

We also recommend provision of funds for the development of *in situ* stress-measuring equipment for absolute stress, and for the variation of stress with time in deep holes, at the rate of $1 million per year for at least five years. Within the continental drilling organization there should be a panel—representing industry, government, and the universities—to coordinate and encourage this development, which is critically needed for adequate understanding of earthquake physics.

A special urgent problem demanding increased research effort is that of reservoir-induced earthquakes. Deep holes are necessary at a number of damsites, monitored for elements such as stress and pore pressure, both before and after reservoir filling, if we are to gain an understanding of this perplexing phenomenon. Federal agencies concerned with dam building, such as the U.S. Bureau of Reclamation, the U.S. Army Corps of Engineers, and the Tennessee Valley Authority, have an obligation to support research in this area, perhaps on a joint basis, because it does not seem appropriate to charge the research costs to the particular dams that are chosen for extensive instrumentation.

VIII. New Initiatives

Any credible level of funding for earthquake-hazard research will necessarily limit dedicated drilling to a few crucial locations—at present the San Andreas fault zone and the New Madrid and Charleston regions. Thus targets of opportunity must be identified and exploited as much as possible if we are to broaden our understanding of seismicity throughout the United States. Dedicated drilling outside California is likely to be done on an *ad hoc* basis related to programmatic needs and ensuing funding. Selection of other specific locations from the vast sample of the North American continent should be made on an individual basis, as targets of opportunity are identified.

We feel that funding of the order of $5 million annually for the first year and $10 million thereafter could be profitably spent for *add-on experiments*, provided that opportunities are widely advertised to the scientific community well in advance of primary drilling.

It is clearly desirable to have a wide range of input into a national drilling project from the scientific community at large. New ideas and proposals should be solicited regarding the following:

- Measurements that will clearly delineate fundamental mechanisms
- Applications of measurements that are already being carried out
- Improved instrumentation

IX. Summary of Recommendations

1. A major instrument-development program is required to develop more satisfactory means of measuring absolute stress at depth and to monitor its change with time.

2. A mechanism should be established to ensure rapid and thorough dissemination of information on drilling intentions to interested scientists.

3. A panel to evaluate holes of opportunity for add-on experiments should be set up by the Department of Energy for evaluating the scientific potential of DOE holes. The same procedure should be considered by other agencies that have drilling programs.

4. A major effort should be made, primarily through targets of opportunity, to obtain absolute stress data from holes throughout the United States, including areas of apparent low seismicity.

5. The drilling of holes within and close to major fault zones, particularly the San Andreas fault of California, should be expanded, with a 10-km hole envisaged in 1984, if results remain encouraging.

6. All major drilling dedicated to the earthquake-hazard program should be coordinated with the U.S. Geological Survey, in recognition of its lead mission-oriented agency role as specified in the Earthquake Hazards Reduction Act of 1977.

Appendix E
Drilling, Logging, and Instrumentation: Technological Limits and Development Needs

E DRILLING, LOGGING, AND INSTRUMENTATION

The objectives of the Drilling, Logging, and Instrumentation Panel were as follows:

- To establish the requirements for drilling, logging, and instrumentation (DLI) based on the programmatic objectives of the scientific panels;
- To assess the capabilities of currently available drilling, logging, and instrumentation technology and current DLI technology development programs to fulfill these requirements; and
- To identify additional research and development activities in drilling, logging, and instrumentation that will be needed to achieve the objectives of the Continental Scientific Drilling Program.

I. Scientific Objectives and Drilling, Logging, and Instrumentation Requirements

A. Basement Structures and Deep Continental Basins

The scientific objectives of this program will address many topics with highly varied drilling requirements. Tectonic and geodynamic questions, such as the early origin of the granitic crust, the identification of buried Precambrian structures, studies of the extent of the edge of the continent in the Precambrian, Paleozoic, and Mesozoic studies of the petrology of the deep crust, and the vertical movements of arches and basins have been identified. While some of these questions can be solved with moderate-depth drilling (< 5 km), others will require ultradeep drilling (10 km or more).

The primary concern of scientists addressing these questions is the collection of representative rock samples from great depth. Good quality, uncontaminated samples for petrological and physical property analyses and radiometric age dating will help solve many of the questions of the origins of the deep crust and the location of ancient sutures and continental margins. Oriented rock samples are needed for studies of the metamorphism and paleomagnetics of the deep crust. Sedimentary rock samples, with precisely located

146

depths, may provide evidence on ages and the paleoenvironment from which subsidence and uplift rates of basins and arches may be calculated.

Advanced logging techniques are desired to measure physical properties that will be critical for the scientific interpretation of the drill holes and samples. Downhole instrumentation (e.g., seismometers, geophones, *in situ* stress, and long-term temperature instruments) is desired to aid in the interpretation of the drilling data and to measure the physical environment.

Several projects involving basement structure and deep basins have been identified that will require deep drilling. A 10-km or deeper probe into the oldest Precambrian basement of Minnesota to study the primordial (3 b.y.) crust is planned. In addition to depth, other drilling requirements are imposed by the following: high temperatures (probably 300°C); casing and cementing problems arising from possibly permeable zones, and loss of circulation; and the physical dimensions of the hole, which must be compatible with downhole instrumentation.

Another project involves a 10-km or deeper probe into the Gulf of Mexico Basin, to basement of unknown character. A prognosis of this probe includes drilling through overpressured shale saturated with hot saline brines; effective blowout prevention and hole control will be needed. Drilling through the Louann salt will require saline muds and casing of both the shale and the salt to prevent flow-in. Temperatures of 300°C might be anticipated at 5-km depths.

Many of these data requirements duplicate those of other research programs. It is hoped that the Continental Scientific Drilling Program management structure will design an effective system for the measurement of all possible parameters in these expensive holes.

B. Thermal Regimes of the Earth's Crust

The program objective is to determine the means of heat transfer into geothermal reservoirs from more primary heat sources. It is necessary to determine the relative roles of heat generation, conductive, convective, and transient processes of heat transfer beneath and around a well-established and well-known geothermal system. Prime candidate sites are The Geysers (California), the Salton Sea geothermal field (California), and Valles caldera (New Mexico). At each of these areas, drilling to meet the objectives will probably involve depths where the temperature exceeds the melting point of at least some of the minerals in a rock. This places a heavy temperature-capability burden on drilling equipment and on sample recovery and *in situ* measuring equipment. *The minimum DLI requirement posed as part of the scientific objective is that representative samples of rocks and formation fluids be obtained and that measurements of temperatures and fluid pressure be made.*

A subsidiary objective is to obtain information on heat generation and transfer from drill holes of opportunity. The DLI requirements related to this objective are to develop means for determining temperatures, heat transport mechanisms, and relevant rock properties with as little impact as possible on the drilling effort.

C. Mineral Resources

The objective of this program is to improve understanding of fundamental principles regarding mineral deposits by obtaining three-dimensional knowledge of deposit formation mechanisms. Information regarding source, transportation (migration), and accumulation of selected minerals is sought. Types of systems to be studied range across numerous mineral assemblages and various development stages of deposit formation. Interests include both currently inactive systems, where history must be inferred, and active systems which, it is presumed, represent current and future accumulations. Mechanisms to be studied are primarily hydrothermally based, with temperatures ranging up to and possibly including magma source levels, and some unusual accumulation phenomena, which cover special environments such as varying pH and oxidation state.

With the exception of the high-temperature environments and the special requirements related to unusual phenomenon studies (such as natural-state core and fluid samples), conventional drilling, well completions, and well-testing technology appear adequate to meet the needs of the Panel on Mineral Deposits. The logging and instrumentation requirements cover conventional petroleum industry technology, include some new technology developed for nuclear test monitoring, and probably will involve some currently conceptual well-testing techniques that will require validation. Both single- and multiple-well testing will be needed to achieve an understanding of the three-dimensional source transport deposit interactions.

The greatest technological challenge, aside from high temperature technology, is to obtain natural-state information by drilling, sampling, and testing with minimum effect on the formation and formation fluids. Techniques must be evolved to achieve unaltered, oriented measurements on and of materials that may be reactive with conventional drilling fluids and hardware.

D. Earthquakes

In the study of earthquakes a primary objective is to obtain data on the crustal distribution of absolute stress and its variation with time. Heat flow data are also important. The value of drill holes in this work is well established.

On the basis of current work, several needs in methods and instrumentation were identified. Despite the success of downhole methods of stress measurement, a need exists for improved methods of determining *in situ* stress as a function of time and depth, especially nondestructive methods.

Another anticipated general requirement is the extension of this capability to deeper holes and higher temperatures. Thus whatever downhole instrumentation is in use now or developed in the future should include the capability to operate at higher temperatures.

Specific current needs include downhole apparatus for overcoring undisturbed (pressurized) core samples, and an improved televiewer. Further, for downhole apparatus used in long-term monitoring, improved cables and cable-handling equipment are needed.

II. Status of Drilling and Logging Technology Relevant to Continental Scientific Drilling Program

A. Drilling

The accomplishment of the scientific objectives of the Continental Scientific Drilling Program will depend to a great extent upon the capabilities of the drilling industry to perform drilling operations at the desired sites and to the desired depths.

Standard techniques will allow the drilling of boreholes at temperatures to 200°C and 10-km depths. Drilling to depths of 16 km is thought to be possible, if temperatures are less than 200°C. Drilling at temperatures between 200° and 350°C is currently possible on a somewhat experimental basis, as an extension of the present capability. Drilling and completing wells at temperatures in excess of 350°C are limited by the lifetime of elastomers, the unavailability of suitable drilling fluids, the lack of suitable cement for setting casing, and the loss of strength in steel tubular goods at higher temperatures.

Table E-1 illustrates the limits of currently available technology in the various temperature ranges. Drilling at temperatures below 200°C represents the present capability for most of the items noted, although some problems are encountered with cementing, elastomers, and sampling in this temperature region.

At temperatures of 350°–1100°C, many unsolved problems exist. Conventional tools fail, and concepts that will permit drilling into progressively hostile environments are neither obvious nor plentiful. Much of the drilling in the Continental Scientific Drilling Program will be at temperatures below 350°C, but as that temperature is approached and exceeded, drilling and maintaining the hole constitute a major technical challenge.

TABLE E-1   Limits of Conventional Well Technology

	Drilling Equipment	Fluids	Wellbore	Sampling
Standard practice, 200°C	Available technology	Available technology	Cement	Coring
Experimental and geothermal, 350°C	Bits, tools, tubulars, wellhead	Liquids, gas, aerated fluids, corrosion	"Casing" packers	Samples, logging, stress state
New regime, 350°- 1100°C	No technology exists	Cooling (possible), new technology needed	No technology exists	No technology exists
Penetration into melt	Experiments pending	Technology exists	Technology exists	Technology exists

Some conclusions may be drawn about the present capabilities of drilling as they apply to the Continental Scientific Drilling Program (they may need to be revised as sites and scientific programs become more specific):

- Drilling equipment and trained personnel are now able to conduct drilling operations to depths of 10 km. While drilling equipment designed for depths greater than 10 km exists, drilling is limited to operations with downhole temperatures of less than 200°C.
- For temperatures of less than 200°C, standard drilling and completion practices may be used; therefore existing tools and techniques are usable for a large portion of the footage to be drilled. The upper portion of all holes may be conventionally drilled.
- Current drilling research programs and industrial efforts will supply needs for drilling at temperatures between 200° and 350°C.
- Above 350°C, concepts and methods for drilling and completing wells do not currently exist. Extension of current rotary drilling techniques to temperatures above 500°C appears to be extremely difficult. Most of the current drilling and well-completion capabilities fail in the range of temperatures from 350°-1100°C. Up to 500°C it may be possible to make improvements in materials that would allow drilling and completion of wells. But it is unlikely that current methods can be extended beyond 500°C. Therefore it will be necessary to create new concepts for drilling and completion. These concepts will be more revolutionary than evolutionary.

B. Logging

The availability of tools for logging wells is shown in Table E-2. Tool diameters and existing temperature and pressure capabilities are listed.

III. Current Activities in Development of Improved Drilling and Logging Technology

A. Industry

1. Information Recovery

a. Coring. Recently, a hybrid core bit based on a four-cone JOIDES bit design has been developed that incorporates Stratapax® (a registered trademark of the General Electric Co.) inserts for the critical core-shaping operation. This hybrid design will improve hard-rock coring by extending the bit life and therefore will increase core recovery per bit.

b. Logging. Logging technology is currently capable of working to temperatures of 250°C, and for short-term use to 300°C. The development of the higher-temperature instrumentation is being conducted jointly with government-funded research labs.

c. Hole surveying and downhole steering. Recently, two companies have provided high-temperature gyroscopic borehole-surveying tool services. With subsequent logging-cable improvement, these systems should be capable of hole surveys at temperatures up to about 350°C (formation temperature). These companies are also improving the heat shielding for their magnetometer and other surveying tools used as downhole steering devices. These steering tools will require a large diameter (5.5 in., or 140 mm) drill string.

2. Drilling

a. Optimized drilling. Several companies are actively developing telemetry systems for the measurement of downhole drilling conditions, to optimize the bit penetration rate and/or time on bottom. These methods will allow measurements such as the force applied to the bit, the bottom-hole temperature and pressure, and the pressure drop across the bit during drilling.

Several measurement-while-drilling (MWD) systems are in the field for testing.

b. Bits. The major advances being made in conventional tricone bits are in improved materials, higher-temperature metallic bearing seals, and improved

TABLE E-2 Commercial Well Logging and Sampling Devices

Type of Log or Sampler	Measurable Parameters	Principal Parameters	Outside Diameter, inches (cm)	Pressure, psi (MPa)	Maximum Temperature, °C
Nonfocused resistivity	Resistivity of materials	Lithology, invasion	3 3/8 (8.6)	20,000 (138)	232
Induction	Conductivity of materials surrounding probe	Lithology, porosity, water quality	3 3/8 (8.6)	25,000 (172)	260
Focused resistivity	Resistivity of formations	Lithology, porosity, water quality	4 (10.2)	25,000 (172)	260
Caliper	Average hole diameter	Lithology, fracture, location, correction of other logs	2 5/8 (6.7)	25,000 (172)	260
Deviation	Angle and direction of hole drift	Location of hole	3 3/8 (8.6)	20,000 (138)	204
Natural gamma	Total gamma intensity	Lithology	2 5/8 (6.7)	25,000 (172)	260
Gamma-gamma	Gamma rays scattered from a source	Bulk density, porosity	2 5/8 (6.7)	25,000 (172)	260
Neutron	Hydrogen density around instrument	Water-saturated porosity	2 5/8 (6.7)	25,000 (172)	260
Pulsed neutron	Neutron capture cross section	Fluid identification porosity	1 11/16 (4.3)	16,500 (114)	150
Nuclear magnetism*	Number and state of hydrogen nuclei	Effective porosity permeability	5 1/2 (14.0)	20,000 (138)	177
Acoustic velocity†	Transit time of elastic waves	Lithology, porosity	3 3/8 (8.6)	25,000 (172)	260
Temperature	Temperature	Water flow, heat flow	1 1/2 (3.8)	20,000 (138)	204
Fluid sampler	(Pad sealed against rock, opens formation to sampler chamber)	Sample of in situ water and formation pressure, permeability	5 (12.7)	20,000 (138)	177
Sidewall sampler	(Charge forces sample bullet into formation)	Laboratory analyses of core sample	4 3/8 (11.1)	20,000 (138)	220
Natural gamma ray* spectroscopy	Uranium, thorium, potassium concentrations	Same	3 3/8 (8.6)	20,000 (138)	260
Downhole flowmeter	Fluid flow rates	Same	1 11/16 (4.3)	15,000 (103)	260

*Limited availability.
†Digital recording on magnetic tape is available.

manufacturing processes (automated alignment and welding). Recently developed artificial diamond cutters (e.g., Stratapax®) may improve the cutting performance of conventional bits. An experimental bit using Stratapax® is being developed jointly by Sandia Laboratories and industry. Because of the higher rpm capability of drag bits, they may have considerable application in deep hot drilling when used with downhole motors.

In addition, work is being done to improve the elastomer seals in bearing bits to extend their operating temperature beyond 250°C.

3. Drill String

Investigation of exotic alloys for use in deep, high-pressure, corrosive environments is under way. This work, initiated specifically for oil field applications, will facilitate the development of special tubulars for the more severe scientific drilling environment foreseen in the Continental Scientific Drilling Program.

4. Drilling Fluids

Considerable progress has been made in the past two years in the development of high-temperature drilling muds. One company is currently marketing a 315°C drilling mud. In high-temperature, unfractured, crystalline rocks, water with a small amount of corrosion inhibitor (ammonium sulfite) is an excellent drilling fluid.

At present, high-temperature foams are being developed by several companies for use in subpressured hydrothermal drilling environments. Previously, air (with its associated corrosion problems) has been the only suitable drilling fluid.

5. Blowout Preventers

Work is being done to develop elastomers with higher operating temperatures. Rotating heads (kelly packers) with higher operating pressures are being investigated.

Downhole blowout preventers are currently being developed for conventional oil field use. With proper modification they may have application to scientific drilling programs.

6. Downhole Motors

There has been a recent upsurge in industrial development of downhole motors, particularly turbines designed for high-temperature drilling conditions. Several companies are developing their own high-temperature turbines. In the very near future, downhole motors will be available that will provide faster drilling rates and extended drill string fatigue life. Pertinent to several phases of the Continental Scientific Drilling Program will be the ability to deviate (directionally drill) holes by using these high-temperature turbines. In addition, work is being done on electric downhole motors.

7. Completion

a. Casing. Present work being done with exotic alloys for critical oil field applications will facilitate the development of casing for the Continental Scientific Drilling Program. In addition, there is a continuing industry program to upgrade tubular steel metallurgy for critical applications, such as H_2S and CO_2 environments and conventional deep drilling.

b. Cement. Industry is working on a project funded by the Department of Energy to increase the temperature at which cement can be used to $350°C$. Equipment to allow greater selectivity in cement placement is also under development. This, coupled with developments in open-hole packers, may serve as a suitable substitution to cement-only systems.

c. Packers. An essential part of any Continental Scientific Drilling Program will be the ability to measure open-hole pore pressure and permeability, the *in situ* state of stress, and effective rock mass elastic moduli. It will also be necessary to sample uncontaminated pore fluids. For most or all of these measurements, a high-temperature open-hole packer system will be required. However, if the typical oil field requirement for retrievable packers is relaxed to *drillable* packers, several existing open-hole steam injection packer designs appear to be almost directly applicable to open-hole formation testing and hydraulic fracturing operations associated with this program.

d. Wellheads and hangers. Investigation is under way in the development of wellhead equipment for high-temperature change (and thus high-tensile-load change) applications.

B. Department of Energy

1. High-Temperature Drilling-and-Completion-Technology Development

a. Bits

- Roller cone bits. The Division of Geothermal Energy (DGE) of the Department of Energy is conducting research programs directed at improving the performance of roller-cone bits for geothermal applications. Existing seal materials and lubricants are currently being tested to determine the best available materials for high-temperature applications. A seal tester with a capability of 5000 psi (35 MPa) at $300°C$ has been developed for this purpose and is available for use by interested parties. In addition, a lubricant tester with a similar capability has been constructed. These devices allow rapid screening of candidate materials. Results from this testing program are being made available to the bit-manufacturing companies.

Improvements in unsealed bits are being sought through the substitution of better materials at key points in the bit design. For example, materials that retain hardness at high temperature have been selected and incorporated into the bearing cones of an air-cooled bit. Several bits using these selected materials have been fabricated and have been tested in an air-drilling laboratory test facility that provides simulated downhole conditions of temperature and pressure.

- Stratapax® bits. Improvements in roller-cone bit design have the potential for increasing the temperature capability of these tools to approximately 350°C. At higher temperatures, different bit designs, e.g., drag bits, will be required. Research directed at utilizing synthetic diamond cutters on drag action bits is currently under way, sponsored by two DOE divisions, the DGE and the Division of Oil, Gas, Shale, and *In situ* Technology (DOGSIST). Since the Stratapax® cutters are capable of high-temperature operation (approximately 700°C), they are attractive for geothermal applications; however, suitable techniques for attaching these cutters to the bit must be developed.
- Stratapax® coring bits. Hybrid coring bits, which use roller cones with Stratapax® inserts for the core-shaping operation, have been built by private industry and tested.
- Downhole replaceable bits. The cost of drilling wells is increased by the periodic requirement to replace a worn bit. In deep holes in hard rock where trip times are long and bit life is short, tripping costs become excessive. DOE/DGE is currently funding the development of a downhole replaceable drill bit. The bit uses diamond-cutting surfaces mounted on the links of a continuous chain. Laboratory tests of this bit indicated that the bit has a significantly increased lifetime over that of conventional bits, while maintaining comparable penetration rates.

b. Drilling and completion fluids. The use of drilling fluids at temperatures in excess of approximately 200°C is extremely difficult. Viscosity and gel-strength considerations require the use of thinners in the mud formulation of higher temperatures. Lost-circulation control additives are frequently rendered ineffective at temperatures above 200°C. This fact requires either the development of high-temperature muds or substitution of an aerated fluid for the mud. However, well control can be a severe problem when drilling with aerated fluids.

- High-temperature fluid development. DGE is currently funding the development of high-temperature drilling muds, high-temperature drilling foams, and laboratory facilities for measuring the properties of drilling fluids under simulated borehole conditions. A sepiolite-based mud using brown coal as a thinner has been formulated and demonstrates usefulness to temperatures of 250°C. A present capability assessment of the use of drilling foams in high-temperature drilling applications is now under way.

- High-temperature fluid-testing facilities. A university laboratory facility for measuring the rheological properties of drilling mud under simulated borehole conditions of 3,000 psi (21 MPa) and 275°C has been constructed. The design of a mud test flow loop capable of 20,000 psi (140 MPa) and 350°C has been initiated, under joint DOE/industry funding.

- High-temperature completion fluids. Completion fluids include those used in drilling through producing formations and in cleaning holes before they are completed. These fluids are generally designed to be nondamaging to the formations of interest. The high-temperature characteristics of fluids commonly used in oil field completions are not defined. DGE has recently initiated efforts to assess the high-temperature performance limits of these fluids and to develop new fluids as needed. As a first step, the static aging characteristics of existing fluids at elevated temperatures are being determined.

c. Downhole motors. Downhole motors are used for directional drilling and for increasing the penetration rate in straight-hole drilling. DGE is currently funding the development of downhole, fluid-driven turbines capable of operating at a formation temperature of 250°C for 200 hours with fluid cooling. In addition, research is under way to design, fabricate, and test a bearing-and-seal package for use in downhole motors. Toward this end, a facility for testing the performance of candidate seals under conditions of temperature (300°C), pressure (5000 psi, or 35 MPa), and rotation has been constructed. A 25 ft (8 m) well bore simulator for testing the entire bearing and seal package at simulated borehole conditions has also been fabricated. These facilities are available for use by interested parties.

d. Well design. The design of geothermal well completions to ensure safety and long life is of paramount importance to the economic viability of geothermal energy. DGE is currently funding research directed at understanding the design requirements for high-temperature well completions. This effort requires the analysis of thermal stresses in the casing and threaded connections, an understanding of primary cementing technology, an assessment of the performance of high-temperature cements, an appreciation of corrosion and scaling problems, and a knowledge of the high-temperature strength properties of tubular goods.

e. High-temperature packers. Packers are required for measuring the open-hole pore pressure and permeability, the *in situ* state of stress, the effective rock mass elastic moduli, and downhole pressures in hydraulic fracturing operations. Currently available oil field packers generally use elastomers to provide the sealing capability required. Elastomers are generally not useful at temperatures above 200°C. DGE is funding the testing of available packers developed by private industry in the Hot Dry Rock Program. In addition, new packer designs that incorporate features that minimize the stress on the elastomeric seals are being developed under DGE funding.

f. High-temperature materials development. DGE and DOGSIST are funding efforts directed at characterizing the performance of materials in the geothermal environment and at developing new materials as needed. Primary programmatic thrusts are the development of high-temperature cements and elastomers and understanding and controlling the high-temperature corrosion problem.

- High-temperature cements. Cement is used in virtually all well completions. The use of cements at high temperatures requires the addition of retardants during emplacement, and the lifetime of cements under exposure to high temperatures and brines is short with commonly used formulations. DGE is funding research directed at developing high-temperature cements capable of providing extended life at temperatures to 350°C. This work involves the selection, formulation, and testing of polymer-based cements. Static aging tests of candidate formulations at geothermal temperatures, and exposure to brine, are being conducted.

- Elastomers. The use of elastomers is pervasive throughout the drilling industry. For example, they are used in blowout preventers, packers, and bit seals. DGE is funding work directed at improving the service life of elastomers, with a goal of 24 hours of service life at 260°C. Screening of existing elastomers for tensile strength and chemical stability at elevated temperatures is under way. Based on these measurements, new chemical formulations will be proposed to provide operation at higher temperatures.

  DOGSIST is funding work on the development of high-temperature elastomers used in sour-gas environment. Approaches under study include the coating of available elastomers with organic compounds, e.g., Teflon, to increase service life. Successful static sealing tests have been conducted at 275° for 24 hours, using a coating of the chemical paralene C cross-linked to the elastomer viton.

- Corrosion studies. Equipment used in drilling and completing geothermal wells is exposed to an extremely corrosive environment. In many instances, air is used as a drilling fluid, and at high temperatures it causes excessive corrosion of tubular goods. Completion equipment is exposed to high-temperature brines. DGE is funding research directed at corrosion phenomena in iron-based alloys. The program involves tests of many materials in high-pressure equipment where a wide variety of brine chemistries can be studied. The purpose of this work is to clarify corrosion factors that influence materials selection for geothermal power plants and drilling equipment.

g. Molten-rock drilling technology. The Division of Basic Energy Sciences (DBES) of the Department of Energy is currently funding feasibility investigations on extracting energy from magma bodies. As part of this research, investigations on high-temperature drilling systems are being conducted.

- Magma simulation facility. A facility for heating rock to 1600°C and 60,000 psi (410 MPa) has been constructed for simulating magma bodies. This facility is to be used for conducting microbit studies at various rock temperatures and pressures and for laboratory testing of drilling techniques for melt penetration.
- Rock mechanics. An associated program involves an investigation of rock properties at and near melt temperatures. Measurements of brittleness and ductility of selected specimens are being performed. These measurements will provide the basis for the design of bits used for drilling rock at very high temperatures.
- Lava lake drilling. Attempts to drill into the Kilauea Iki lava lake (Hawaii) were made in 1976. Penetration of the melt to a depth of 1 m was achieved. A new drilling system using an insulated drill string and high-temperature drag bit has been designed and fabricated. Testing of this system at Kilauea Iki is planned in 1979.

2. Logging and Instrumentation Technology

a. High-temperature logging tools. Sandia Laboratories, under DGE sponsorship, manages the Geothermal Logging Instrumentation Technology Development Program, whose primary objective is to develop the technology to log high-temperature geothermal holes. Emphasis is placed on technology initially for 275°C applications; extension to 350°C is envisioned. Some developments would have capability in excess of 500°C.

Areas of potential interest to the Continental Scientific Drilling Program include the development of the following:

- Passive thin- and thick-film electronic components (275°–350°C)
- High-resolution temperature flow and pressure-logging tools (275°C)
- Active electronic circuits with ultimate capability greater than 500°C
- Downhole refrigerator for providing cooling to electronic packages
- Formation temperature tool using nuclear techniques
- Support capability for running experimental and prototype tools in a hole

b. Pressure coring system. A system designed to obtain core samples and maintain them under bottom-hole pressures is being designed with funding from the Department of Energy's Bartlesville Energy Research Center.

While improvements in mechanical design are an important feature of this project, an equally important feature is the use of a noninvading coring fluid in the barrel. This fluid is designed to protect the core from the drilling mud and to provide minimum invasion of the core so that low residual hydrocarbon saturation levels may be detected.

c. Fracture detection. The location and orientation of both artificially produced and natural fractures intersecting boreholes is of critical importance to the DGE programs. The borehole televiewer (BHTV), which is an acoustic scanning device that can provide a visual record of $360°$ circumference of the borehole, has been available for a number of years but is not being currently used because of its complexity, unreliability, and limited applicability. The only BHTV in use today is operated by the U.S. Geological Survey (USGS). DGE is pursuing a cooperative program among the USGS, Sandia (the BHTV licensee), and contractors, for upgrading both the reliability and the temperature capability of the BHTV. A reasonable temperature objective is $275°C$.

d. Borehole fluid sampler. DGE has funded the development of a borehole fluid-sampling device, which is capable of operating at downhole conditions of $200°C$ and 6000 psi (41 MPa). The tool is capable of collecting two fluid samples at different depths.

e. Downhole pH probes. The long-term monitoring and control of geothermal energy production require both plant process and downhole instrumentation. Emphasis will be placed initially upon plant process instrumentation, to be followed by development of instrumentation for *in situ* measurements. A critical need is for pH measurements, and the development of a pH probe is included in a "request for proposal" recently released by Battelle's Pacific Northwest Laboratory as part of the DGE program.

f. Borehole-borehole communication (signaling). Borehole-to-borehole communication by acoustic or other techniques may be of limited interest; however, extensive efforts in this area have been made by the Los Alamos Scientific Laboratory (LASL) in the Hot Dry Rock Program. The Department of Energy's Waste Management Program is also supporting limited technology development in this area. The interest in these programs is on determining geological features and structures in the earth between the two boreholes.

g. Thermal conductivity and heat flow probes. DGE has been supporting the development of a thermal conductivity probe and a heat flow probe. The initial data evaluation is encouraging.

3. Directional Drilling

The ability to drill, locate, and maintain directional holes is critical to the recovery of many energy resources and may have application to the Continental Scientific Drilling Program. DOGSIST is funding several research efforts directed at improving directional drilling capability.

a. Electrodrill®. The Electrodrill® (a registered trademark of the General Electric Company) concept involves the use of a downhole electric motor for rotating the bit at high rpm. A 50-hp system has been built and tested, and a 285-hp system is under construction. The use of wires to supply power to the motor also provides a channel for data transmission; hence measurements during drilling may be possible.

b. Downhole turbines. DOGSIST is funding development of small-diameter (5-3/8 in., or 136.5 mm) turbines, through the Morgantown Energy Research Center, for the purpose of drilling directional holes in coal seams.

c. Mud-pulse telemetry system. Of critical importance in directional drilling is a knowledge of hole deviation and bit position. DOGSIST, in cooperation with private industry on a cost-sharing basis, has funded the development of a downhole telemetry system that uses pressure modulation of the circulating mud column to transmit directional information from the bit to the surface, on a real-time basis. A directional sensor inside a Monel drill collar is used to determine positional information and then to encode this information by modulating a poppet valve in the circulating mud column. The information is then decoded at the surface. This system has been tested in several offshore wells and has been highly successful.

C. U.S. Geological Survey

The U.S. Geological Survey does not have a broad instrumentation development program; rather, specific instruments are being developed to support well-defined research programs. Of particular interest to the Continental Scientific Drilling Program are the following:

- A hydraulic fracturing system for *in situ* stress determinations. The system consists of a borehole televiewer, an improved packer design to permit repeated inflation and deflation without withdrawal from the hole to redress the tools, and electronic pressure transducers to sense the pressure in the fractured interval as well as to detect leaks around the package.
- A velocity system. This combines, via a microprocessor, the usual full-wave sonic-log data with the natural fracture information (borehole televiewer) and stress state (hydraulic fracturing method) to deliver velocities and attenuation that are close to the borehole.

IV. Advances in Instrumentation for Geoscience Applications

Recent advances in geoscientific instrumentation are as follows:

A. Downhole Logging Measurements

1. Cable Construction for High-Temperature Borehole Logging

Single-conductor fiberglass insulation is available for use to 325°C. Polypropylene insulation is available for use to 290°C. Multiple-conductor (seven-conductor) polypropylene and Teflon are useful to 250°C. Research is under way for high-temperature elastomers to extend cable temperature range.

2. Communication (Telemetry) Methods

A pulse code modulation system is now available that permits transmission of up to 14 signals on a single conductor line with temperatures up to 290°C for several hours.

3. Pressure-Measuring Sensors and Instrumentation

This instrumentation is available from two or more sources, using ultraprecise quartz crystal and ceramic sensor elements. One of these will operate to 275°C.

4. Borehole Gravimeter

This device will detect subsurface density anomalies and measure densities of rock, both in open-hole and cased-hole environments. It should also be useful for earthquake prediction or mineral detection because it also determines density in horizontal strips. Resolution of density changes up to 100 m or more from the borehole may be achieved.

5. Cryogenic Magnetometers

This instrument measures magnetic susceptibilies of rock formations to great accuracy and produces a continuous lithology log.

6. Nuclear Magnetism Logging

This technique locates and quantifies the presence of free fluids in rock formations. A good measure of rock permeability may be obtained in carbonate formations. The presence of producible fluid and measurement of volume fluid in permeable formations of all types may be indicated.

7. Carbon-Oxygen Ratio Log

This technique measures radioactive decay of fluids and distinguishes between oil (hydrocarbons) and water; it may be used for detection of radioactive minerals and measurements of heat productivity.

8. Differential Temperature Log

This device measures the temperature differential between two points on the borehole instrument. It is used in locating fluid or gas flow entering the borehole through fractures and for determining the depth of the top of the cement column and cement channeling in cased holes behind the casing. This device is available from three or four commercial sources.

9. Absolute-Temperature Measurement

Measurements of the temperature of the borehole fluid provide indications of the formation temperatures and reasonably accurate temperature gradients with depth. Considerable additional research for measurement of the absolute temperature is now being conducted, using thermalized neutron methods.

10. Thermal Conductivity of Formations

This method uses a laser flash and is now under development.

11. Overshot Strain Gauge (Overcoring)

This has been developed on at least one prototype and should still be available.

12. Absolute-Stress Measurement With Depth and Time

a. Absolute stress with depth. The present system for this measurement has poor accuracy. An overshot core would solve this problem.

b. Measurement of magnetic permeability (borehole magnetometer) and dielectric constant (dielectric log) changes may be useful. These services are available.

13. Acoustic Detectors

Detection of high-frequency acoustic emission indicates rock bursts before failure. Research is in progress.

14. Air Gun or Vibroseis Stacking Experiment for High Gain

High-precision seismic velocity monitoring is now in the research stage.

15. Downhole Dilatometer

Downhole dilatometer equipment is now in the research stage.

16. Downhole Tilt

Downhole tilt measurements to determine consistency or dependence of tilt on depth are available.

17. Radar and Sonar Transillumination

This area encompasses radar frequencies of 30–1000 MHz and sonar frequencies of 100 kHz with pulse durations of 1–10 μs. Also associated are backscatter surveys and tomography (to determine formation structure in the vicinity of the hole) involving multiple-travel-path transillumination. Limitations are about 85°C temperature and 100 psi (0.7 MPa), since the boreholes are generally about 100 ft ($\approx$ 30 m) in depth.

18. Triaxial Borehole Magnetometer

Available from a foreign source, it is useful for measuring formation permeability and for determining lithology. A major oil company owns an alternating current induction tool for magnetic permeability measurements, and one has been built for a government agency.

19. Cation Exchange Log and Induced Polarization Measurements

One commercial source has built a cation exchange tool, and several industrial sources provide service using induced polarization. The benefits of these devices should be evident to the mining interests.

20. Optical Borehole Televiewer

This is not to be confused with the acoustic televiewer. A possible use for the *optical* version for fracture detection is in air-drilled holes, but it could also be extended to the infrared range and used for temperature scans. An alternative, using low-light-level television, has been suggested as a possibility.

B. Surface Instrumentation Advances

1. Magnetotelluric Magnetometer

Latest variations in modern usage include multiple magnetometers combined with telluric data.

2. Cryogenic Magnetometer

This is an ultrasensitive magnetometer, operating in a cryogenic atmosphere. Principal uses are for mapping geologic anomalies such as salt domes and fault zones, for mineral detection and identification, and for location of geothermal hotspots.

3. Long-Offset Crustal Scale Vibroseis

A seismic method using geophone spacings of from 0.5 to 50 km for geophysical exploration is available.

4. Use of Color in Data Presentation

This will add a fourth dimension to a three-dimensional plot. One application is in the integration of data of diverse types.

5. Computer Graphics

This can provide a powerful means for analyzing multidimensional data.

6. Stress-Strain Measurements

This method uses strain gauges with tiltmeters.

7. Heat Flow Measurements

There is a method for the separation of geothermal heat flow from solar heat flow in shallow holes of only 1–2 m in depth. A thermopile is used.

8. Seismic Stacking on Horizontal Levels

This is a method to obtain Poisson's ratio and the amplitude of the reflecting horizon. This is akin to tomography and involves reflection stacking and transmission stacking.

9. Remote Sensing

Computer-enhanced photographic images of world-wide areas are available from the Earth Resources Orbiting Satellite. Even more important is the procurement of digital-taped image data that may be easily enhanced by computer processing at the user's discretion.

10. Spectral Imagery

Principal use of these data is for the detection and identification of shallow mineral deposits and uses such as the pinpointing of polluted water and stream bodies.

11. Airborne Magnetics

Airborne magnetics are used for geological fault and mineral location as well as for Curie point determination.

12. Low-Level Emanation Determination

This is equipment for measurement of mercury concentrations to 1 ppb. It is also used for helium and radon emanation measurements.

13. Passive Seismic Detection

This method involves the following:

- Identification and location of ambient seismic noise
- Location of small earthquakes
- Identification of P- and S-wave delays from distant earthquakes

14. Mud Logging

The sampling of fluids going into the well compared to similar data obtained from the outflow of the mud column provides chemistry, lithology, particle size, and degree of radioactivity. Also, these data may be used to provide a rough determination of true formation temperature.

V. Technology Development Needs

A. Basement Structures and Deep Continental Basins

Many of the specific problems of deep-drilling technology have been addressed in the report on magma-hydrothermal drilling and instrumentation (Varnado and Colp, 1978). Development needs exist for drilling at the 300°C and higher temperatures expected at 10 km and greater depths. The deep-basin drilling will involve contact with hot saline brines, which causes concern about corrosion of casing and drilling equipment. These developmental problems are also addressed by Varnado and Colp (1978).

Some of the drilling proposed involves high-pressure shale zones in deep basins. Drilling experience with this problem indicates that it can be handled by conventional drilling and casing. Pressures can be controlled by existing blowout protection systems (improvements in these systems are being developed).

Drilling through salt in the Gulf of Mexico is proposed. Experience in the drilling of salt for nuclear waste storage indicates that this can be done mechanically or with solution drilling.

The deep casing called for in some holes will require tapering, but the diameter at bottom may be kept to a reasonable size, even at a depth of 10 km. This does not present a problem for downhole instrumentation, which is commonly available at 2.5- to 3.0-in. (6.3–7.6 cm) diameters.

Developmental logging is needed for ultradeep drilling. Velocity logs and downhole gravimeters are available, but at present cannot operate at temperatures above 260°C. Paleomagnetometers for downhole logging are not currently available, but they should be developed.

Sampling technology requires some development for ultradeep drilling in the area of sidewall corers, which are currently limited to temperatures below 220°C because of the explosive charges. Electronics in the gyroscope and magnetic compass attachments for oriented cores are limited to less than 250°C. The capability of these devices should be upgraded.

B. Thermal Regimes of the Earth's Crust

Developmental requirements relate to the following:

1. The Need to Obtain Samples That Are Diagnostic of Processes Involved in Heat Transfer

a. Sampling must be done at high temperatures and in potentially corrosive environments.

b. Sampling must not contaminate the samples taken. In particular, very low levels of trace elements may need to be recognized, in addition to common materials such as water, steam, and solids in solution. It may be necessary to maintain samples at bottom-hole temperatures and pressures, until critical analyses are made.

2. The Need to Make Novel Physical Measurements

a. Temperatures of the undisturbed rock, along with thermal properties (including heat capacity, thermal conductivity, and/or diffusivity) need to be measured. While equipment and techniques for measuring these properties have been under development, it is not apparent how these measurements can be made successfully in a deep, hostile borehole environment.

b. Calculation of mass transfer of heat requires determination of at least two of three sets of quantitative measurements on rocks in place: fluid pressures and/or pressure gradients *in situ* in the pore structures, fluid mobility or viscosity, and rock permeability. Again, these determinations must be made in a high-temperature corrosive environment, but it may also be necessary to make these measurements in rocks that are normally considered impermeable. Detection of fluid movement in microfractures is a particularly challenging development problem. Although these approaches are also under study, the instrumental solution to the measurement requirement remains to be designed.

3. Communication of Data From Downhole Sensors to the Surface

a. Present cables are operable to temperatures to $350°C$, if specially designed. They can be upgraded to $600°-700°C$ using special materials, particularly where digital transmission on a single cable is used.

b. At melting point temperatures, it may be necessary to use downhole data recorders. Because of the difficulty in obtaining information, the priority of measurements and minimization of redundancy are important.

4. Conventional Borehole Surveying Equipment

Conventional borehole surveying equipment should be upgraded in terms of survivability at high temperatures.

5. Quantitative Means for Measuring Heat Production

There is a need for the development of quantitative means for measuring heat production. There is equipment for spectral gamma-ray logs, so the need is to develop reliable means for converting such data to heat production information.

6. Petrophysical Studies

Petrophysical studies in unusual rock types are needed. Evaluation of many types of geophysical logs requires knowledge of the relationship between measured quantities and the rock property desired. Such relationships are well-known for oil-bearing rocks at lower temperatures but virtually unknown for other rock types and at higher temperatures.

C. Mineral Resources

Most of the needs related to fundamental principles of the proposed mineral deposit program may be met by currently available drilling, logging, and instrumentation technology. The unmet needs are associated with elevated temperatures that have been discussed previously in connection with high-temperature drilling operations. Unique unmet needs for the mineral deposit program are associated with sampling and with drilling operations in abnormally corrosive environments. Special downhole instrumentation may also be required, as follows:

1. Corrosion

New materials and techniques will be required for drilling in high-temperature, corrosive brine operations. Chloride brines appear to be of particular concern, although H_2S gives problems too.

2. Sampling

The need to obtain and/or measure uncontaminated, unaffected samples of fluids and solids is fundamental to the mineral deposit program. Therefore special new drilling fluid and sampling technology for natural state core recovery and for obtaining fluid samples with unaltered chemical compositions, down to and including trace level components, will be required. This new sampling concern must also extend to operations at high temperature.

3. Instrumentation

This area of need will be defined as specific mineral deposit programs are undertaken. Already known needs include *in situ* permeability, fracture density and orientation, *p*H (at high temperature and pressure and in ambient brines), sulfide activity, and Eh. Mapping technology via acoustic or electromagnetic methods will also need improvement because of the new environments being studied.

D. Earthquakes

The user of drill holes in long-term monitoring of such phenomena as seismic activity, stress variations, and heat flow requires long-life cables and a system for efficiently handling such cables in emplacement, servicing, and recovery operations. High-temperature versions of such cables are also needed.

Data acquisition systems, primarily digital, and including high-temperature downhole components, are also needed in connection with this monitoring effort. (For earthquake research it is not anticipated that data are required during drilling. However, such a requirement will apply for other parts of the program.) Although adequate data acquisition systems are in current use by service companies, they are apparently not readily available for procurement and long-term use by others. A similar situation seems to apply to the cable and cable-handling systems discussed above.

A primary objective of earthquake-related research is detailed knowledge of the crustal distribution of absolute stress and its variations with time. The methods now in use or proposed for this work can, in principle, provide part of this information. Most of these methods require drill holes.

One method now being used provides measurements of pressure variations associated with hydrofracturing. An important aspect of this method is the prefracture and postfracture mapping of the cracks in the selected length of the drill hole wall. The televiewer used for this mapping work should be designed to be more easily handled and should provide higher resolution in both area and intensity.

In the class of destructive measurements of *in situ* stress there are at least two other methods to be considered—overcoring and laboratory measurements of core samples. Both appear to require some hardware development. The apparatus that has been used for downhole overcoring has not performed adequately. A pressurized core sampler that would maintain the sample at its *in situ* pressure state is also needed.

There is also a need for instrumentation with which to monitor variations in stress. A downhole stress gauge should be developed that can easily be inserted into a hole to measure temporal changes of the magnitude and directions of the principal stresses. One such device might consist of a soft-wall pressure vessel that could be expanded against the hole. Changes in the cross section of the hole (and therefore the vessel) could be measured optically. The absolute value of the principal stresses might be obtained by hydrofracturing.

Looking to the future, it is clear that most of the earthquake-related, downhole experiments and measurements will be made in deeper holes as they become available. Thus it is important to develop high-temperature versions of any applicable apparatus that would not be expected to function properly in the higher-temperature environment.

VI. Conclusions and Recommendations

A. Conclusions

1. Current drilling and logging technology is adequate to fulfill the drilling requirements of the Continental Scientific Drilling Program (CSDP) down to depths of 10 km or up to temperatures of 200°C, whichever occurs first.

However, improved fluid and solid sampling techniques will be required to support many scientific investigations.

2. Present research and development activities by government and industry will ultimately supply drilling needs at temperatures up to 350°C.

3. Practically all scientific objectives of the CSDP will eventually require drilling at temperatures of 500°C and higher and at depths of 10–15 km. There is currently no capability for drilling, logging, and completing wells in such an environment, and little research is being conducted.

4. There are needs in several phases of the CSDP to develop new techniques for collecting uncontaminated fluid and solid samples.

5. Earthquake studies require the development of remote (nondestructive) techniques for *in situ* stress measurements.

6. Materials suitable for use in high-temperature, highly corrosive environments are required in all phases of the CSDP, except possibly earthquake studies.

7. Borehole instruments for strain and tilt measurements are not available.

8. Successfully obtaining data in support of the CSDP via add-on experiments in new holes will require that investigators make measurements in holes of opportunity on a minimum-response-time basis and with minimum interference with the drilling operation.

B. Recommendations

1. A coordinating panel for drilling and instrumentation research and development should be established to ensure that results of ongoing research activities are available to all participants in the CSDP. This panel should include representatives from government, national laboratories, private industry, and universities. Consideration should also be given to including participants from other countries.

2. Program needs should be widely disseminated to provide an opportunity for private industry to participate in the research and development activities.

3. The existing technology development program funded by the Department of Energy should be expanded to include development of drilling, completion, and logging techniques capable of operating to 500°C and beyond. Of critical importance is the development of elastomers and seals, drilling tools, and completion techniques for use at these temperatures.

4. Early efforts in the technology development program should be directed at understanding the scientific objectives of the CSDP, defining development research and development tasks that are required to accomplish these objectives, planning the implementation of these tasks (including time and cost estimates), and recommending a procedure for accomplishing the required research and development tasks.

5. Development of the following instruments should be initiated:

- A bottom-hole sampler for obtaining uncontaminated fluid and solid samples from the borehole.
- A system for performing *in situ* stress measurements on a nondestructive basis.
- Borehole strainmeters and tiltmeters.
- *In situ* neutron-activation tools optimized for detection of specific minerals.

Appendix F
Implementation of a National Continental Drilling Program for Scientific Purposes

F IMPLEMENTATION OF A NATIONAL CONTINENTAL DRILLING PROGRAM FOR SCIENTIFIC PURPOSES

I. Introduction

The scientific community has long recognized the need for a program in crustal studies making more effective use of information from existing and new drill holes to address fundamental scientific questions on the physical and chemical properties of the crust, the structure of the crust, and active processes operating in the crust. The scientific questions and institutional mechanisms for addressing them effectively were the subject of the Workshop on Continental Drilling for Scientific Purposes convened at Ghost Ranch, New Mexico, June 10–13, 1974 (Shoemaker, 1975) and of a follow-on study (also entitled "Continental Drilling") by the Federal Coordinating Council for Science and Technology through its Panel on Continental Drilling of the FCCSET Committee on Solid Earth Sciences (FCCSET, 1977). These two reports, considered in the light of the development of ongoing federal drilling programs, constitute a major part of the background of the Workshop on Continental Drilling for Scientific Purposes, July 1978.

Subsequent to the above-mentioned studies, drilling activities by federal agencies, principally the Department of Energy, the Department of the Interior, and the Department of Defense, have increased rapidly. In fiscal year 1979, expenditures by federal agencies will exceed one-half billion dollars.

Thus, *de facto*, the United States has a major federal drilling effort, but it does not have a national continental drilling program for scientific purposes. Currently, drilling by the federal government is carried out within the many programs of mission-oriented agencies. Clearly, it is in the public interest to maximize the scientific return from this very large investment of public funds. The design of a program of continental scientific investigation, using drilling as a powerful tool, must be predicated primarily on priorities developed from scientific considerations. It will be constrained by pragmatic concerns of available opportunities within mission priorities and within available funding, but the fundamental concern of the scientific program must be the scientific significance of the proposed effort.

To help ensure the realization of these scientific goals, we recommend establishment of a mechanism to enhance communication and interaction between the scientific community and the federal agencies so as to do the following:

- To provide scientific and technological advice to the agencies
- To provide information to the scientific community on drilling programs of the federal agencies and on opportunities for research
- To recommend scientific objectives for augmenting the mission-oriented programs and identify drilling projects for scientific purposes
- To effect interagency coordination of drilling activities for scientific purposes
- To establish a data management system
- To establish a coordinated system of libraries for cores, samples, and logs
- To encourage international exchange of relevant data derived from continental drilling in other countries
- To evaluate periodically the effectiveness of the coordinated program in enhancing the quality and quantity of the scientific and technological product

We considered alternative mechanisms to accomplish the objectives and reviewed the recommendations in the 1975 and 1977 reports. However, because of progress made by the Department of Energy and the Department of the Interior in establishing mission-oriented drilling programs over the last several years, we concluded that it is necessary to reconsider the form of a Continental Scientific Drilling Program as suggested in 1975 or of an Interagency Continental Drilling Program as recommended in 1977. *In our judgment, the very substantial multiagency drilling program now in progress will benefit from broad-based scientific advice.*

Accordingly, we suggest that the several federal agencies engaged in significant drilling activities establish and provide funding support for a Board on Continental Drilling for Scientific Purposes. The supporting federal agencies should interact directly with the board through liaison representatives. The board should interact with the scientific community through appropriate panels, workshops, and published reports. Such a board would greatly facilitate interagency coordination. Accordingly, an interagency committee on drilling is a possible mechanism for improving coordination and maximizing benefits. Critical to the successful operation of this mechanism is action by each participating federal agency to establish internally a central office or clearinghouse to coordinate drilling activities within and between agencies.

II. Role of Proposed Board on Continental Drilling for Scientific Purposes

To accomplish the scientific objectives of the Continental Scientific Drilling Program, we suggest that interested federal agencies take the initiative to establish a Board on Continental Drilling for Scientific Purposes composed of scientists and engineers representing appropriate disciplines. Three-year rotating memberships are recommended, with nominations drawn from academia, state and federal agencies, industry, and scientific and professional societies. Geoscientists from the Congressional Research Service and the Office of Technology Assessment should be considered for appointment as Congressional Observers, along with other scientific and technical affiliates. Care should be taken to maintain representation from all relevant fields.

Figure F-1 shows a suggested structure for the board and how it would interface with supporting federal agencies that have drilling programs. The activities of the board should be conducted by a permanent staff of sufficient size. It is important that the board interact with boards, committees, and panels with related responsibilities.

Much of the work of the board would require input from specialists in the relevant disciplines. Such input might be best received and coordinated by established panels of specialists. The number of panels, as well as the membership on each panel, should remain flexible to meet the particular needs at any given time. Use of the workshop format to evaluate a particular scientific question has proved effective.

The board should have the following functions and responsibilities:

- *Advise federal agencies on needs for drilling for scientific purposes.* The board would receive from its panels recommendations for drilling activities needed to meet particular scientific objectives. These recommendations would be reviewed periodically by the board and incorporated into a proposed program of continental drilling for scientific purposes. Of necessity, the program would be dynamic, reflecting the changing needs for information perceived by the scientific community. The program plan would be reviewed periodically with the participating federal agencies, to identify opportunities to meet the particular scientific objectives. Where possible, such objectives would be incorporated into the existing drilling programs of the agencies, either through minor program modification or through the addition of objectives to a proposed drilling program with additional incremental funding.
- *Advise agencies on needs for new technology.* The identification of needed drilling for scientific purposes will develop the need for new technol-

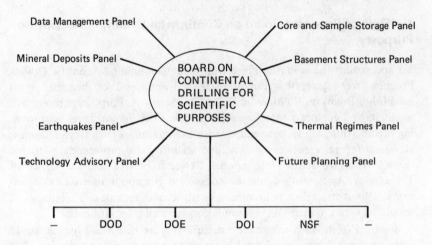

FIGURE F-1   Proposed Board on Continental Drilling for Scientific Purposes and interface with federal agencies engaged in drilling programs.

ogies in drilling, sampling, and borehole measurements. The board would have the responsibility for advising the participating agencies of needs for new technologies to accomplish the program's scientific objectives more effectively.

- *Provide for the dissemination of information from the federal agencies to the scientific community.* An important responsibility of the board would be to provide for the dissemination of information from the federal agencies to the scientific community. Such communication is important to ensure an effective, continuing flow of information on drilling for scientific purposes.

- *Identify needed funding for scientific research involving drilling and related corollary research not included in agency drilling program plans.* The program of drilling for scientific purposes can be met in part through a more effective coordination of the current drilling programs of the federal agencies. However, requirements will be identified that cannot be met by existing programs and will therefore require new funding. The board will be expected to identify the need for such funding and develop the appropriate justification for its support in cooperation with the federal agencies.

- *Provide for interaction with other national and international agencies and organizations to promote scientific studies of the continents.* Continental studies are currently being conducted by many agencies and organizations around the world. A fuller dissemination of information derived from such studies would be mutually beneficial. The board would investigate mechanisms to improve the dissemination of information from the Continental Scientific Drilling Program to the world scientific community and to receive, in return, information from comparable investigations in other parts of the world.

III. Role of Federal Agencies

The federal Departments of Energy, Defense, and the Interior are currently spending this year at least one-half billion dollars on drilling designed to satisfy the needs of specific federal programs. Without impairing program objectives, we urge that these large federal programs in drilling be augmented by a basic science-oriented effort aimed at a better understanding of the fundamental composition, structure, and processes of the continental crust. Such a thrust will result in improved knowledge of the continental crust and subsurface processes and will contribute to the long-term effectiveness of their mission programs.

Where possible, we urge that a component of scientific drilling be incorporated within each mission program. This may be accomplished at a small fraction of the cost of the mission drilling program if the mission agencies and the National Science Foundation assume the following responsibilities:

- Establish a focal point for drilling programs within each agency
- Establish a suitable mechanism for interagency coordination of the scientific effort
- Keep the board currently and fully informed of federal drilling activities and plans
- Interact with the board in an effort to optimize scientific return
- Fund appropriate add-on science-related research activities to mission drilling programs and provide the necessary budgetary flexibility to support purely scientific drilling
- Involve and support the broadest possible academic, federal, and industrial scientific communities in the scientific and technical aspects of continental drilling
- Cooperate with other federal agencies to emphasize and realize the broad scientific potential associated with mission drilling programs
- Implement and publicize a system of information, data banks, and sample repositories that are mutually compatible and accessible to the general scientific community

IV. Interagency Coordination

To ensure coordination in implementation of the Continental Scientific Drilling Program and to provide an appropriate interface with the board, it is important that the participating federal agencies interact to achieve the following:

- Day-to-day coordination between the agencies in program planning, preparation, and implementation
- Early agreement on responsibility for core and sample storage and standards
- Optimum utilization of agency capabilities and facilities in cooperative efforts and projects
- Coordinated interaction with the board so that the scientific community, through the board, may be informed about agency operations
- Cooperation, as required, to include all federal agencies wishing to participate in continental drilling for scientific purposes
- Collection and dissemination of information pertaining to drilling from all federal agencies involved in significant drilling activities on land (including the continental margins)
- Assessment and resolution of legal and regulatory problems associated with drilling operations
- Sharing of downhole research support and site surveys
- Appropriate arrangements for assumption of responsibility for core repositories with general guidance from the board

We have recommended that the operating functions necessary for a Continental Scientific Drilling Program be assigned to and coordinated among the participating federal agencies. We recognize, however, that as the scientific drilling program evolves and more holes are drilled for purely scientific purposes, it may be desirable to centralize operating functions in a separate management entity for scientific drilling.

V. Information and Data Management

A successful Continental Scientific Drilling Program will generate large quantities of important new data. In addition, there are currently vast amounts of data from diverse sources that are relevant to a Continental Scientific Drilling Program. Some existing subsurface data sources, such as very large accumulations of petroleum industry well data files, are readily accessible; others, such as mining industry data, are largely held in proprietary files and may not be as readily available to the scientific community. Abundant data from current drilling operations are available from industry, state, and federal operators. Industry is drilling in new geographical and geological frontiers for energy and mineral resources. Several federal agencies have active drilling programs that provide pertinent data for improved understanding of the continental crust.

These data must be evaluated, disseminated, and combined with information to be derived from the proposed drilling plan to maximize results of the scientific drilling program.

We recommend that the agencies establish an information and data management system to support the scientific drilling program. This should be a coordinated agency effort to provide efficient processing, storage, and dissemination of usable data to the scientific community and to the user groups. The agencies should adopt formats and codes that would permit common access and reference to data stored in agency files. The board, through a panel on data management, should advise the agencies on scientific data needs.

The agency information and data management system should do the following:

- Adopt a standardized data system, including a common index and access system, to meet the needs of the agencies and the scientific community
- Work to standardize existing formats and codes and increase compatibility between appropriate existing agency and institution data bases
- Develop a rapid and effective dissemination system
- Evaluate the capabilities of existing data systems
- Monitor current and proposed public and private drilling activity, to identify holes that could be used for acquisition of scientific data and help to publicize the objectives of the scientific program
- Inform the scientific community as rapidly as possible about agency drilling plans

VI. Sample and Core Repositories

The success of a Continental Scientific Drilling Program depends upon an effective archival system for the cores, samples, and other borehole information derived from drilling. Holes drilled for purely scientific purposes present problems that are different from those of scientific additions to mission-oriented holes with existing or augmented scientific objectives. The system therefore must do the following:

- Provide the scientific community with ready access to all materials obtained through the program
- Ensure proper preservation and storage of samples of cores, cuttings, and fluids in a mode that provides for ready retrieval
- Establish a good working index
- Disseminate catalogs of available material
- Integrate materials produced in the continental drilling program with the existing system of private, state, and federal well sample and core libraries

- Develop policies and procedures to provide for circulation of working collections of material for analyses and experiments

We recommend that the archival function be assumed by the participating federal agencies. Considering the current level of drilling activity, there is an urgency to make the decision.

The board should provide advice to the archival agency on organization alternatives and funding levels. Items for consideration include the following:

- The establishment of a unified, coordinated system of regional archives
- Establishment of standard operating procedure for participating agencies
- The provision of federal support to existing state archives to upgrade service and ensure conformance with established procedures

VII. Review and Evaluation

We recommend that at the conclusion of an appropriate period, there be a formal and comprehensive assessment of the performance of the system, addressed to success in these areas:

- Providing interaction between the scientific community (academic scientists, industry scientists, professional and scientific societies, and state agencies) and the federal agencies
- Effecting both intra-agency and interagency coordination
- Enhancing the scientific dimensions of mission-oriented drilling through direct program expansion or dual-purpose drilling
- Generating support for wholly scientific drilling projects
- Establishing efficient data management and core and sample archival systems

This assignment should be carried out by an independent *ad hoc* review committee.

Appendix G
Organization of the Workshop

Advance planning for the workshop and conduct of the workshop itself were guided by a Steering Committee: Elburt F. Osborn (*Chairman*), Charles L. Drake, Howard R. Gould, George A. Kolstad, Jack E. Oliver, Eugene M. Shoemaker, and Leon T. Silver.

On behalf of the Steering Committee, Eugene Shoemaker was responsible for ensuring that the report of the workshop was completed by the end of the workshop. The main activities of the workshop were conducted by six panels, listed below. The reports of these panels appear in Appendixes A-F.

Panel	*Chairman*
Basement Structures and Deep Continental Basins	Jack E. Oliver
Thermal Regimes of the Crust, Particularly Those Related to Hydrothermal and Magmatic Systems	Robert W. Decker
Mineral Resources	Brian J. Skinner
Earthquakes	Clarence R. Allen
Drilling, Logging, and Instrumentation: Technological Limits and Development Needs	Harold M. Stoller
Implementation of a National Continental Drilling Program for Scientific Purposes	Howard R. Gould

Agency Liaison Representatives to the Workshop on Continental Drilling for Scientific Purposes:

John G. Heacock, Office of Naval Research
George A. Kolstad, Department of Energy
John F. Lance, National Science Foundation
Gordon A. Swann, U.S. Geological Survey

Appendix H
Background Documents
for the Workshop

As part of the advance planning for the workshop, the following background papers were prepared and distributed to workshop participants prior to the beginning of the workshop.

A Background on Crustal Dynamics and Related Subjects   Charles Drake
B Scientific Directions in Solid Earth Studies, and How Continental
   Drilling Can Contribute to the Solution of Problems
    1 Basement Structures and Deep Continental Basins   Jack Oliver
    2 Thermal Regimes of the Crust—Particularly Those Regimes
       Related to Magmatic and Hydrothermal Processes   Robert Decker
    3 Mineral Resources—Fundamental Principles   Paul Barton
    4 Direct Measurements in Boreholes of Properties of Fault
       Zones   Barry Raleigh
C Papers Discussing Scientific Holes and Projects
    1 Piggy-backing on the Michigan Basin Hole (preprint of
       Journal of Geophysical Research article by N. Sleep
       and L. Sloss, with introduction   Laurence Sloss and
       by Sloss)   Norman Sleep
    2 A Proposed Drilling Program at Washburn Hot Springs,
       Yellowstone National Park   Hubert Barnes
    3 Several Topics   Kenneth Deffeyes
       a Shapes of Hydrothermal Systems
       b Hydrothermal Ore Deposition During Late Diagenesis
       c Pressure-Volume-Temperature Data on Pore Fluids
       d Novel Drilling Techniques
    4 Precambrian Deep-Hole Targets in Minnesota   Matt Walton
    5 Coordination of Continental Drilling Program and Federal
       Hot Dry Rock Development Program   Morton Smith

References

Arth, J. G., and G. N. Hanson. 1975. Geochemistry and origin of the early Precambrian crust of northeastern Minnesota. *Geochim. Cosmochim. Acta 39*:325–362.

Balk, R. 1930. Structural survey of the Adirondack anorthosite. *J. Geol. 38*: 289–302.

Bayley, R. W., and W. R. Muehlberger, compilers. 1968. Basement rock map of the United States, exclusive of Alaska and Hawaii. U.S. Geological Survey Two-Sheets.

Björnsson, S., and P. Einarsson. 1974. Seismicity in Iceland. In *Geodynamics of Iceland and the North Atlantic Area*, L. Kristjansson, ed. D. Reidel, Boston, Mass.

Boyer, R. E. 1962. Petrology and structure of the southern Wet Mountains, Colorado. *Geol. Soc. Am. Bull. 73*(9):1047–1069.

Brander, J., R. G. Mason, and R. W. Calvert. 1976. Precise distance measurements in Iceland. *Tectonophysics 31*:193–206.

Buddington, A. F. 1939. Adirondack igneous rocks and their metamorphism. *Geol. Soc. Am. Mem. 7.*

Burke, K., L. Delano, J. F. Dewey, A. Edelstein, W. S. F. Kidd, K. D. Nelson, A. M. C. Sengor, and J. Stroup. 1979 (in press). *Rifts and Sutures of the World*. Geophysics Branch, Earth Science Applications Div., NASA/ Goddard Space Flight Center, Greenbelt, Md.

Carder, D. S. 1945. Seismic investigations in the Boulder Dam area, 1940–1944, and the influence of reservoir loading on earthquake activity. *Bull. Seismol. Soc. Am. 35*: 175–192.

Carpenter, A. B., M. L. Trout, and E. E. Pickett. 1974. Preliminary report on the origin and chemical evolution of lead and zinc rich oil field brines at central Mississippi. *J. Econ. Geol. 69*:1191–1206.

Carter, J. L. 1977. Comparison of ultramafic and mafic xenoliths from Kilbourne Hole and Potrillo maar. In *Second International Kimberlite Conference Extended Abstracts*. Geophysical Laboratory, Carnegie Institution of Washington, Washington, D.C.

Chapin, C., and W. Seager. 1975. Evolution of the Rio Grande rift in the Socorro and Las Cruces areas. In *Las Cruces Country*. New Mexico Geological Society Guidebook, 26th Field Conference. Pp. 297–321.

Chapin, C., R. Chamberlin, G. Osburn, D. White, and A. Sanford. 1978. Exploration framework of the Socorro geothermal area, New Mexico. *N. Mex. Geol. Soc. Spec. Publ. 7*:114–129.

Craddock, C., H. M. Mooney, and V. Kolehmainen. 1970. Simple Bouguer gravity map of Minnesota and northwestern Wisconsin. Minn. Geol. Surv. Misc. Map Ser. M-10. Scale 1:1,000,000.

Daigniers, M., and G. Vasseur. 1978 (in press). Détérmination et interprétation du flux géothermique à Bournac (Haut-Loire). *Geophys. Rev.*

Deffeyes, K. 1969. The electron macroprobe, a new instrument for logging deep sea cores. Abstract. *Geol. Soc. Am. Abstr. Program.* p. 264.

Dorman, J., J. L. Worzel, R. Leyden, N. Crook, and M. Hatziemmanuel. 1972. Crustal section from seismic refraction measurements near Victoria, Texas. *Geophysics 37*(2):325-336.

Engel, A. E. J., and C. G. Engel. 1958. Progressive metamorphism and gravitization of the major paragneiss, northwest Adirondack Mountains, New York. *Geol. Soc. Am. Bull. 69*:1369-1414.

Ervin, C. P., and L. D. McGinnis. 1975. Reelfoot rift: Reactivated precursor to the Mississippi Embayment. *Geol. Soc. Am. Bull. 86*:1287-1295.

FCCSET. 1977. *Continental Drilling.* Recommendations of the Panel on Continental Drilling of the FCCSET Committee on Solid Earth Sciences. 19 pp.

Flawn, P. T. 1965. Basement—Not the bottom but the beginning (abstract). *Am. Assoc. Petrol. Geol. Bull. 48*:524-525.

Fridleifsson, I. B. 1973. Petrology and structure of the Esja Quaternary volcanic region, southwest Iceland. Ph.D. thesis, Oxford. 208 pp.

Fridleifsson, I. B. 1977. Distribution of large basaltic intrusions in the Icelandic crust and the nature of the layer 2/layer 3 boundary. *Geol. Soc. Am. Bull. 88*:1689-1693.

FUSOD, 1977. *The Future of Scientific Ocean Drilling.* Report by an *ad hoc* Subcommittee of the JOIDES Executive Committee. Seattle, Wash. 92 pp.

Goldich, S. S. 1968. Geochronology in the Lake Superior region. *Can. J. Earth Sci. 5*:715-724.

Goldich, S. S. 1972. Geochronology in Minnesota. In *Geology of Minnesota: A Centennial Volume.* P. K. Sims and G. B. Morey, eds. Minnesota Geological Survey. Pp. 27-37.

Goldich, S. S., and C. E. Hedge. 1974. 3,800 myr. granitic gneiss in southwestern Minnesota. *Nature 252*:467-468.

Goldich, S. S., C. E. Hedge, and T. W. Stern. 1970. Age of the Morton and Montevideo gneisses and related rocks, southwestern Minnesota. *Geol. Soc. Am. Bull. 81*:3671-3696.

Goldich, S. S., J. L. Wooden, G. A. Ankenbauer, Jr., T. M. Levy, and R. U. Suda. 1976. Precambrian history of the Morton-New Ulm reach of the Minnesota River valley. Abstract. In *22nd Annual Institute on Lake Superior Geology.* Minnesota Geological Survey. P. 22.

Haimson, B. C., and B. Voight. 1977. Crustal stress in Iceland. *Pure Appl. Geophys. 115*(1/2):153-190.

Hast, N. 1973. Global measurements of absolute stress. *Phil. Trans. Roy. Soc. London 274*:409-419.

Hildenbrand, T. G., M. F. Kane, and W. Stauder, S. J. 1977. Magnetic and gravity anomalies in the northern Mississippi Embayment and their spatial relation to seismicity. U.S. Geol. Surv. Misc. Field Invest. Map 914.

Hoffer, J. M. 1975. The Aden-Afton basalt, Potrillo volcanics, southcentral—New Mexico. *Tex. J. Sci. 26*(3-4):380-390.

Hubbert, M. K., and W. W. Rubey. 1959. Role of fluid pressure in mechanics of overthrust faulting. I. Mechanics of fluid-filled porous solids and its application to overthrust faulting. *Geol. Soc. Am. Bull. 70*:115-166.

Jonsson, J. 1967. The rift zone and the Reykjanes Peninsula in Iceland and mid-ocean ridges, S. Björnsson, ed. *Soc. Sci. Island. 38*:142–148.

Kjartansson, G. 1960. Geological map of Iceland. Sheet 3, SW Iceland, Menningarsjodur, Reykjavik.

Klein, F. W., P. Einarsson, and M. Wyss. 1973. Microearthquakes on the mid-Atlantic plate boundary on the Reykjanes Peninsula in Iceland. *J. Geophys. Res. 78*:5084–5099.

Klein, F. W., P. Einarsson, and M. Wyss. 1977. The Reykjanes Peninsula, Iceland, earthquake swarm of September 1972 and its tectonic significance. *J. Geophys. Res. 82*:865–888.

Kolstad, C. D., and T. R. McGetchin. 1978. Thermal evolution models for the Valles caldera with reference to a hot-dry-rock geothermal experiment, *J. Volc. Geotherm. Res. 3*:197–218.

Kristjannsson, L. (ed.). 1974. Geodynamics of Iceland and the North Atlantic area. D. Reidel, Boston, Mass. 324 pp.

Lachenbruch, A. H., and J. H. Sass. 1977. Heat flow in the United States and the thermal regime of the crust, in *The Earth's Crust*, J. G. Heacock, ed. *Am. Geophys. Union Geophys. Monogr. Ser. 20*:626–675.

Lambert, I. B., and K. S. Heier. 1967. The vertical distribution of uranium, thorium and potassium in the continental crust. *Geochim. Cosmochim. Acta 31*:377–390.

Leyreloup, A., C. Dupuy, and R. Andriambololona. 1977. Catazonal xenoliths in French Neogene volcanic rocks: Constitution of the lower crust. 2. Chemical composition and consequences of the French Massif Central Precambrian crust. *Contrib. Mineral. Petrol. 62*:283–300.

Luth, W. L., and G. Simmons. 1968. Melting relations in natural anorthosite. In *Origin of Anorthosite and Related Rocks*, Y. W. Isachsen, ed. Mem. 18. New York State Museum Science Service, Albany. Pp. 31–37.

McDougall, I., K. Saemundsson, H. Johannesson, N. Watkins, and L. Kristjansson. 1977. Extension of the geomagnetic polarity time scale to 6.5 m.y.: K-Ar dating, geological and paleomagnetic study of a 3,500-m lava succession in western Iceland. *Geol. Soc. Am. Bull. 88*:1–15.

McGetchin, T. R., and L. T. Silver. 1970. Compositional relations in minerals from kimberlite and related rocks in the Moses Rock Dike, San Juan County, Utah. *Am. Mineral. 55*:1738–1771.

Meyer, C., E. P. Shea, and C. Goddard, Jr. 1968. Ore deposits at Butte, Montana. Graton-Sales Vol. 2. American Institute of Mining, Metallurgical, and Petroleum Engineers, New York. Pp. 1373–1416.

Morey, G. B. 1976. The basis for a continental drilling program in Minnesota. *Minn. Geol. Surv. Inform. Circ. IC-11.*

Morey, G. B. 1977. Geology and mineral resources of east-central Minnesota— Some new perspectives. In *38th Annual Mining Symposium*. University of Minnesota. (Minn. Geol. Surv. Reprint 37.) 24 pp.

Morey, G. B. 1978a. Lower and middle Precambrian stratigraphic nomenclature in east-central Minnesota. *Minn. Geol. Surv. Rep. Invest. 21.* 52 pp.

Morey, G. B. 1978b. Metamorphism in the Lake Superior region, U.S.A., and its relation to crustal evolution. In *Symposium on Metamorphism in the Canadian Shield*. J. A. Fraser and W. W. Heywood, eds. *Geol. Surv. Can. Spec. Paper 78-10*:283–314.

Morey, G. B., and P. K. Sims. 1976. Boundary between two Precambrian W terranes in Minnesota and its geologic significance. *Geol. Soc. Am. Bull.* *87*:141-152.

Morris, H. T., and T. S. Lovering. 1961. Stratigraphy of the East Tintic Mountains, Utah. *U.S. Geol. Surv. Prof. Paper 361.* 145 pp.

Muffler, L. J. P., and D. E. White. 1969. Active metamorphism of Upper Cenozoic sediments in the Salton Sea geothermal field and the Salton Trough, southeastern California. *Geol. Soc. Am. Bull. 80*:157-182.

Nakamura, K. 1970. En echelon features of Icelandic ground fissures. *Acta Natur. Island ii*(8):15.

Nolan, T. B. 1935. The underground geology of the Tonopah mining district, *Univ. Nev. Bull. 29*(5). 49 pp.

Ocean Sciences Board, National Research Council. 1979 (in press). *Continental Margins: Geological and Geophysical Research Needs and Problems.* National Academy of Sciences, Washington, D.C.

Oliver, J., M. Dobrin, S. Kaufman, R. Meyer, and R. Phinney. 1976. Continuous seismic reflection profiling of the deep basement, Hardeman County, Texas. *Geol. Soc. Am. Bull. 87*:1537-1546.

Padovani, E. R., and J. L. Carter. 1977. Aspects of deep crustal evolution beneath south central New Mexico. In *The Earth's Crust*, John Heacock, ed. *Am. Geophys. Union Geophys. Monogr. Ser. 20*:19-55.

Palmason, G. 1971. Crustal structure of Iceland from explosion seismology. *Soc. Sci. Island. 40*:1-187.

Palmason, G. 1973. Kinetics and heat flow in volcanic rift zone with application to Iceland. *Geophys. J. Roy. Astron. Soc. 33*:451-481.

Palmason, G. 1974. Heat flow and hydrothermal activity in Iceland. In *Geodynamics of Iceland and the North Atlantic Area*, L. Kristjansson, ed. Proceedings of the NATO Advanced Study Institute in Reykjavik. D. Reidel, Boston, Mass. Pp. 297-306.

Palmason, G., and K. Saemundsson. 1974. Iceland in relation to the mid-Atlantic ridge. *Ann. Rev. Earth Planet. Sci. 2*:25-50.

Palmason, G., K. Ragnars, and S. Zoega. 1975. Geothermal energy developments in Iceland 1970-1974. In *Proceedings of the 2nd U.N. Symposium on Development and Use of Geothermal Resources.* P. 213.

Raleigh, C. B., J. H. Healy, and J. D. Bredehoeft. 1976. An experiment in earthquake control at Rangely, Colorado. *Science 191*:1230-1237.

Reid, H. F. 1910. The mechanics of the earthquake. Vol. II in *The California Earthquake of April 18, 1906*, A. C. Lawson, ed. Publ. 87. Carnegie Institution of Washington, Washington, D.C. 192 pp.

Reilinger, R., and J. Oliver. 1976. Modern uplift associated with a proposed magma body in the vicinity of Socorro, New Mexico. *Geology 4*:573-586.

Reiter, M., and R. Smith. 1977. Subsurface temperature in the Socorro Peak KGRA [known geothermal resource areas], New Mexico. *Geotherm. Energy Mag. 5*:37-41.

Reiter, M., C. Edwards, H. Hartman, and C. Weidman. 1975. Terrestrial heat flow along the Rio Grande rift, New Mexico and southern Colorado. *Geol. Soc. Am. Bull. 86*:811-818.

Robinson, P. T., W. A. Elders, and L. J. P. Muffler. 1976. Quaternary volcanism in the Salton Sea geothermal field, Imperial Valley, California. *Geol. Soc. Am. Bull. 87*:347–365.

Saemundsson, K. 1974. Evolution of the axial rifting zone in northern Iceland and the Tjornes fracture zone. *Geol. Soc. Am. Bull. 85*:495.

Saemundsson, K., and H. Noll. 1975. K:Ar ages of rocks from Husafell in western Iceland and the development of the Husafell central volcano. *Jökull 24*:40–59.

Sanford, A. 1977. Temperature gradient and heat-flow measurements in the Socorro, New Mexico area, 1965–1968. Geophys. Open-File Rep. 15. Geoscience Department, New Mexico Institute of Mining and Technology. 19 pp.

Sanford, A., A. Budding, J. Hoffman, O. Alptekin, C. Ruch, and T. Toppozada. 1972. Seismicity of the Rio Grande rift. Circ. 120, New Mexico Bureau of Mines and Mineral Resources. 19 pp.

Sanford, A., R. Mott, Jr., P. Shuleski, E. Rinehart, F. Carabella, R. Ward, and T. Wallace. 1977. Geophysical evidence for a magma body in the crust in the vicinity of Socorro, N.M. In *The Earth's Crust*, John Heacock, ed. *Am. Geophys. Union Geophys. Monogr. Ser. 20*:385–403.

Schäfer, K. 1972. Transform faults in Iceland. *Geol. Rundsch. 61*:942.

Second U.N. Symposium on the Development and Use of Geothermal Resources. 1976. U.S. Government Printing Office, Washington, D.C.

Shaw, D. R., and R. L. Parker. 1967. Layered mafic-ultramafic intrusion at Iron Mt., Fremont City, Colorado. *U.S. Geol. Surv. Bull. 1251-A*. Pp. A1–A29.

Shoemaker, E. M., ed. 1975. *Continental Drilling*. Report of the Workshop on Continental Drilling, Ghost Ranch, Abiquiu, New Mexico, June 10–13, 1974. Carnegie Institution of Washington, Washington, D.C. 56 pp.

Sieh, K. E. 1978. Prehistoric large earthquakes produced by slip on the San Andreas fault at Pallett Creek, California. *J. Geophys. Res. 88*(8):3907–3939.

Sigurdsson, H. 1970. Structural origin and plate tectonics of the Snaefellsnes volcanic zone, Western Iceland. *Earth Planet. Sci. Lett. 10*:129.

Silver, L. T. 1968. A geochronologic investigation of the Anorthosite complex, Adirondack Mountains, New York. In *Origin of Anorthosite and Related Rocks*, Y. W. Isachsen, ed. Mem. 18. New York State Museum Science Service, Albany. Pp. 233–251.

Simmons, G. 1964. Gravity survey and geological interpretation, northern New York. *Geol. Soc. Am. Bull. 75*:81–98.

Simpson, D. W. 1976. Seismicity changes associated with reservoir loading. *Eng. Geol. 10*:123–150.

Sims, P. K. 1976a. Precambrian tectonics and mineral deposits, Lake Superior region. *Econ. Geol. 71*:1092–1118.

Sims, P. K. 1976b. Early Precambrian tectonic-igneous evolution in the Vermilion district, northeastern Minnesota. *Geol. Soc. Am. Bull. 87*:379–389.

Sims, P. K., and G. B. Morey. 1973. A geologic model for the development of early Precambrian crust in Minnesota. Abstract. *Geol. Soc. Am. Abstr. Program. 5*:812.

Singewald, Q. 1962. Thorium deposits in the Wet Mountains. *U.S. Geol. Surv. Prof. Paper 300*. Pp. 581–585.

Skinner, B. J. 1976. A second iron age ahead? *Am. Sci. 64*:258–269.

Sleep, N. H., and L. L. Sloss. 1978. A deep borehole in the Michigan Basin. *J. Geophys. Res. 83*:5815–5819.

Sykes, L. R. 1967. Mechanism of earthquakes and nature of faulting on the mid-oceanic ridges. *J. Geophys. Res. 72*:2131.

Tomasson, J., I. B. Fridleifsson, and V. Stefansson. 1976. A hydrologic model for the flow of thermal water in SW Iceland with a special reference to Reykir and Reykjavik thermal areas. In *Proceedings of the 2nd U.N. Symposium on Development and Use of Geothermal Resources.* Vol. 1, pp. 643–648.

Trendall, A. F. 1978. Basins of iron-formation deposition. *Geol. Correl.* (UNESCO) spec. iss.:106–107.

Tryggvason, E. 1973. Seismicity, earthquake swarms and plate boundaries in the Iceland region. *Bull. Seismol. Soc. Am. 63*:1327.

Van Schmus, W. R. 1976. Early and Middle Proterozoic history of the Great Lakes area, North America. *Phil. Trans. Roy. Soc. London 280*:605–628.

Varnado, S. G., and V. L. Colp (eds.). 1978. Report of the workshop on magma/hydrothermal drilling and instrumentation. Rep. Sand78-1365C. Sandia Laboratories, Albuquerque, N. Mex. July 1978.

Walton, M. S. 1977. Crustal structure and seismicity in Minnesota. Abstract. *Trans. Am. Nucl. Soc. 26*:131.

Ward, P. L. 1971. New interpretation of the geology of Iceland. *Geol. Soc. Am. Bull. 82*:2991–3012.

Ward, P. L., G. Palmason, and C. Drake. 1969. Microearthquake survey and the mid-Atlantic ridge in Iceland. *J. Geophys. Res. 74*:665–684.

Warner, L. A. 1978. The Colorado Lineament: A middle Precambrian wrench fault system. *Geol. Soc. Am. Bull. 89*:161–171.

Weber, R. E. 1977. Petrology and sedimentation of the upper Precambrian Sioux Quartzite, Minnesota, South Dakota and Iowa. M.S. thesis. Univ. Minn., Duluth.

Weiblen, P. W., and K. J. Schulz. 1978. Is there any record of meteorite impact in the Archean rocks of North America? In *Proceedings, Lunar and Planetary Science Conference, 9th. Geochim. Cosmochim. Acta, Suppl. 10*:2749–2771.

White, D. E., E. T. Anderson, and D. K. Grubbs. 1963. Geothermal brine wells; mile-deep drill hole may tap ore-bearing magnetic water and rocks undergoing metamorphism. *Science 139*:919–922.

Wold, R. J. 1969. Shallow seismic studies in western Lake Superior. Abstract. In *15th Annual Institute on Lake Superior Geology.* P. 41.

World Data Center A for Solid Earth Geophysics. 1978. *Report SE-14: Directory of U.S. Data Repositories Supporting the International Geodynamics Project.* 40 pp.

Zietz, I., and J. R. Kirby. 1970. Aeromagnetic map of Minnesota. U.S. Geol. Surv. Geophys. Invest. Map GP-725. Scale 1:1,000,000.

Participants

in the
Workshop on Continental Drilling for Scientific Purposes
Los Alamos, New Mexico, July 17-21, 1978

	Affiliation at Time of Workshop
Allen F. Agnew	Library of Congress
Clarence R. Allen	California Institute of Technology
Orson L. Anderson	University of California at Los Angeles
Robert S. Andrews	Office of Naval Research
Ernest E. Angino	University of Kansas
William E. Benson	National Science Foundation
Joseph W. Berg, Jr.	National Academy of Sciences–National Research Council
David D. Blackwell	Southern Methodist University
John D. Bredehoeft	U.S. Geological Survey
Albert L. Bridgewater	National Science Foundation
R. J. Bridwell	Los Alamos Scientific Laboratory
Robert R. Brownlee	Los Alamos Scientific Laboratory
Kenneth Brunot	Department of Energy
B. Clark Burchfiel	Massachusetts Institute of Technology
Kevin C. Burke	State University of New York–Albany
C. Wayne Burnham	Pennsylvania State University
Eugene N. Cameron	University of Wisconsin
James L. Carter	University of Texas at Dallas
Clifton Carwile	Department of Energy
Paul J. Coleman, Jr.	University of California at Los Angeles
John L. Colp	Sandia Laboratories
Robert W. Decker	Dartmouth College
Kenneth S. Deffeyes	Princeton University
Charles L. Drake	Dartmouth College
Robert F. Dymek	Harvard University
Lyman M. Edwards	Dresser Industries
Donald O. Emerson	Lawrence Livermore Laboratory
Anthony W. England	U.S. Geological Survey
Peter T. Flawn	University of Texas at Austin
Frederick Followill	Lawrence Livermore Laboratory
Ingvar B. Fridleifsson	Iceland National Energy Authority
Terrence M. Gerlach	Sandia Laboratories
Howard R. Gould	Exxon Production Research Company
Conway Grayson	Department of Energy

Arthur R. Green	Exxon Production Research Company
John Greenip	Hydrill Company
Anton L. Hales	University of Texas at Dallas
John W. Handin	Texas A&M University
Harry C. Hardee, Jr.	Sandia Laboratories
John F. Harris	Consultant
Pembroke J. Hart	National Academy of Sciences–National Research Council
John H. Healy	U.S. Geological Survey
Hollis D. Hedberg	Princeton University
Grant Heiken	Los Alamos Scientific Laboratory
Julian Hemley	U.S. Geological Survey
John F. Hermance	Brown University
William J. Hinze	Purdue University
Myron K. Horn	Cities Service Oil Company
John M. Jones	Schlumberger Offshore Services
George V. Keller	Colorado School of Mines
George A. Kolstad	Department of Energy
Arthur H. Lachenbruch	U.S. Geological Survey
John F. Lance	National Science Foundation
James F. Lander	National Oceanic and Atmospheric Administration
William C. Luth	Stanford University
Thomas R. McGetchin	Lunar and Planetary Institute
Ian D. MacGregor	Department of Energy
Charles T. Mankin	Oklahoma Geological Survey
H. Jay Melosh	California Institute of Technology
Ralph S. Millhone	Chevron Oil Field Research Company
Glen B. Morey	Minnesota Geological Survey
William R. Muehlberger	University of Texas at Austin
L. J. Patrick Muffler	U.S. Geological Survey
William Ogle	Energy Systems, Inc.
Jack E. Oliver	Cornell University
Steven S. Oriel	U.S. Geological Survey
Michael Oristaglio	National Aeronautics and Space Administration
Elburt F. Osborn	Carnegie Institution of Washington
Elaine R. Padovani	Massachusetts Institute of Technology
William L. Petrie	National Academy of Sciences–National Research Council
Robert E. Riecker	Los Alamos Scientific Laboratory
Carl F. Roach	Department of Energy
John C. Rowley	Los Alamos Scientific Laboratory
Robert F. Roy	University of Texas at El Paso
Alan Ryall	Defense Advanced Research Projects Agency
Charles A. Salotti	University of Wisconsin at Milwaukee
Allan R. Sanford	New Mexico Institute of Mining Technology
Robert N. Schock	Lawrence Livermore Laboratory

Robert E. Sheridan	University of Delaware
Eugene M. Shoemaker	California Institute of Technology
Leon T. Silver	California Institute of Technology
Gene Simmons	Massachusetts Institute of Technology
Brian J. Skinner	Yale University
Norman H. Sleep	Northwestern University
Laurence L. Sloss	Northwestern University
Scott B. Smithson	University of Wyoming
Hartmut A. Spetzler	University of Colorado
Glenn Stafford	Department of Energy
Philip H. Stark	Petroleum Information Corporation
Donald W. Steeples	Kansas Geological Survey
Harold M. Stoller	Sandia Laboratories
Gordon A. Swann	U.S. Geological Survey
Richard Taschek	Los Alamos Scientific Laboratory
Hugh P. Taylor, Jr.	California Institute of Technology
Daniel A. Textoris	Department of Energy
Robert I. Tilling	U.S. Geological Survey
Donald L. Turcotte	Cornell University
Edd R. Turner	Getty Oil Company
Thomas M. Usselman	State University of New York at Buffalo
Samuel G. Varnado	Sandia Laboratories
Rosemary J. Vidale	Los Alamos Scientific Laboratory
Barry Voight	Pennsylvania State University
Matt Walton	Minnesota Geological Survey
Donald E. White	U.S. Geological Survey
Harold A. Wollenberg	Lawrence Berkeley Laboratory
Mark D. Zoback	U.S. Geological Survey